W9-BTM-360

Minerals

Alessandro Guastoni
Roberto Appiani

Minerals

FIREFLY BOOKS

A FIREFLY BOOK

Published by Firefly Books Ltd. 2005

First printing

Publisher Cataloging-in-Publication Data (U.S.)

Guastoni, Alessandro.
Minerals / Alessandro Guastoni ; Roberto Appiani.
Originally published: Italy : Mondadori, 2003.
[256] p. : col. photos. ; cm.
Includes bibliographical references.
Summary: An introduction to the science of mineralogy with an illustrated guide to 288 mineralogical species.
ISBN 1-55407-056-2 (pbk.)
1. Mineralogy, Determinative—Handbooks, manuals, etc. I. Appiani, Roberto. II. Title.
549 /.1 22 QE367.2.G83 2005

Library and Archives Canada Cataloguing in Publication

Guastoni, Alessandro
Minerals / Alessandro Guastoni, Roberto Appiani.
Translation of: Minerali.
Includes bibliographical references.
ISBN 1-55407-056-2
1. Minerals. 2. Mineralogy. I. Appiani, Roberto II. Title.
QE363.2.G8213 2005 549 C2004-907454-7

Published in the United States by
Firefly Books (U.S.) Inc.
P.O. Box 1338, Ellicott Station
Buffalo, New York 14205

Published in Canada by
Firefly Books Ltd.
66 Leek Crescent
Richmond Hill, Ontario L4B 1H1

Art director: Giorgio Seppi
Editorial direction: Tatjana Pauli
Cover design: Chiara Forte
Layout: Studio Grafico Clara Bolduri, Milan
English translation: Jay Hyams
North American consultant: Dr. Robert Martin, Department of Earth and Planetary Sciences, McGill University, Montreal

Most of the specimens in the photographs belong to the collections of the Museo Civico di Storia Naturale di Milano (the Civic Natural History Museum in Milan) and the Federico II Museum of the University of Naples. Some specimens were made available by private collectors, including E. Prato, R. Prato, L. Caserini, C. Albertini, D. Respino, M. Motta, H.P. Frank, A. Carsani, L.M.F. Burrillo, J. Fabre, H. Brukner, B. Ottens, S. Conforti, R. Pagano, E. Ossola, R. Marsetti, E. Ronchi, A. Ferri, and U. Righi, all of whom we thank for their valuable contributions to this book.
A.G. and R.A.

Printed in Spain

CONTENTS

SYMBOLS USED IN THIS BOOK

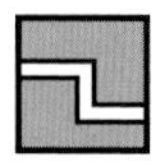

PERFECT CLEAVAGE
the mineral tends to cleave along very regular, planar surfaces

GOOD CLEAVAGE
the planes of cleavage are planar but not perfectly regular

POOR CLEAVAGE
the cleaved surfaces are irregular and no longer planar

INDISTINCT CLEAVAGE
the surfaces of cleavage are conchoidal, with surfaces comparable to those of broken glass

Specific gravity is a number expressing the relative density of a mineral, meaning the relationship between the weight of the mineral and the weight of an equivalent volume of water at a temperature of 39.2°F (4°C).

Hardness is a measure of how difficult it is to scratch a flat surface of a mineral; it is expressed on a scale from 1 to 10.

NOTE TO THE READER

The 288 mineralogical species examined in this book are subdivided into ten **crystallochemical** classes. Many species within these classes are joined in groups (for example, the amphiboles and the zeolites).

The entry for every species gives its common name, chemical formula and crystal system. Each entry then describes the habit of the mineral, the environment in which it forms, the etymology of its name, and any further information useful for the correct identification and description of that species.

Some entries contain two minerals, because both form a series (called a "solid solution") in which the proportion of one or another chemical element can vary.

OPENING-PAGE CAPTIONS

Page 1: Cuprite crystals from the Ray mine, Arizona
Pages 2–3: Liroconite from Cornwall, England
Pages 4–5: Grossular and diopside from the Susa Valley, Piedmont, Italy
Pages 8–9: Beryl from the Codera Valley, Lombardy, Italy
Pages 42–43: Silver with rhodochrosite from Peru
Pages 50–51: Tetrahedrite and realgar from the Binn Valley, Switzerland
Pages 70–71: Fluorite crystals from the People's Republic of China
Pages 76–77: Brookite crystals from Mount Bregaceto, Liguria, Italy
Pages 104–105: Azurite crystals from Sardinia, Italy
Pages 118–119: Boracite from Germany
Pages 124–125: Wulfenite crystals with mimetite from Mexico
Pages 160–161: Elbaite crystals from the island of Elba, Italy
Pages 244–245: Whewellite crystals from the Czech Republic
Pages 248–249: Gold crystals on quartz from Brusson, Valle d'Aosta, Italy

PHOTO CREDITS

All photographs are by Roberto Appiani with the following exceptions:
Page 38b: F. Valoti
Page 222b: B. Ottens
Pages 28a, 29b, 32a, 37b, 38a: F. Pezzotta

FOREWORD

Mineralogy is a complex science with deep roots in chemistry, crystallography and physics. For this reason it is often less than popular with much of the public; for many the subject brings back unpleasant memories, the result of the rigorous and scientific way in which it and its associated disciplines were taught in school.

This has served to make mineralogy the least familiar of the natural sciences, even to nature enthusiasts, who, during their excursions, occasionally encounter specimens of minerals and are almost always fascinated by their find—not just by the geometric perfection of its crystals, but also by its beauty, its shape and color.

This book is not meant to be an exhaustive treatise on mineralogy. It is instead a guidebook designed to reward the efforts of those who wish to increase their knowledge of minerals without running the risk of getting lost in the discipline's more complicated meanders.

Here the skills of expert photographer and collector Roberto Appiani are combined with the talents of Alessandro Guastoni—scientist, collector, museum laboratory director—to create a book that is at once scientifically accurate and aesthetically pleasing, full of wonderfully evocative pictures of mineral specimens. The large selection of entries, covering the most important mineral species, offers a practical, accurate and thorough guide that will prove understandable even to the novice.

Dr. Francesco Demartin
Professor of General and Inorganic Chemistry at the
University of Milan
Crystallographer and passionate collector of minerals

INTRODUCTION

THE SCIENCE OF MINERALOGY

Mineralogy is the science of minerals. To the uninitiated, mineralogy can appear to be a very complicated subject. Many of the disciplines that form its core, such as chemistry, crystallography, physics and mathematics, can constitute forbidding barriers to understanding this fascinating natural science.

This book endeavors to eliminate such barriers and to accompany the reader seeking to enter the world of minerals.

The primary goals of this introduction are to illustrate some of the principal properties of minerals, present the modern tools used in their study, describe the natural environments in which minerals are formed, and explain the criteria by which minerals are classified.

The main section of this book is an illustrated guide to 288 mineral species, all of them illustrated with unique photographs taken by nature photographer Roberto Appiani.

Special thanks go to Dr. Federico Pezzotta, Curator of Mineralogy and Petrography at the Museum of Natural History in Milan, Italy, who wrote the section of this introduction that deals with the environments in which minerals are formed.

Below: *Typical examples of meteorite rocks.*
Bottom: *Natural crystals of sapphire corundum from Sri Lanka.*

Minerals throughout history

Minerals are crystalline substances that are found in their natural state. Minerals are familiar to everyone, as they compose the rocks and mountains around us, as well as the sand on our beaches and the soil in our gardens. Many of the products we use every day are composed of minerals: toothpaste, for example, contains microcrystals of mica, calcite and fluorite, while detergents contain such mineral additives as calcite, dolomite, clays and zeolites. Minerals are components of meteorites and planets, while gemstones are nothing more than rough fragments of crystals, unusually transparent or colorful, that have been cut to emphasize their brilliance and transparency. Minerals have always had great importance in our world; from the dawn of history, each step in mankind's development can be measured by the use of metals. Today, minerals are the principal elements of steel and special alloys, and are integral to electronic and communication devices; they are also used in the space industry and in the manufacture of a great many everyday items.

The science of mineralogy came into being in relatively recent times. In order to understand the scientific criteria that governs its principles, one needs to trace the most important steps along its path over the centuries.

Above: *Specimen of red ocher composed of a mixture of iron hydroxides and oxides.*

The oldest use of minerals is related to art: primitive humans used natural pigments, hematite reds and manganese oxide blacks, to paint the walls of the caves in which they lived. About 5,000 years ago, the Egyptians were making objects from precious metals, using such colored minerals as malachite, lazurite and the emerald variety of beryl.

The first texts to deal with mineralogical subjects were those of the Greek Theophrastus, around 370 B.C., and Pliny the Elder 400 years later. With *Historia Naturalis*, Pliny describes the perfect geometric shapes of crystals, laying the basis for the science of mineralogy. However, it is the German physician and scientist Georgius Agricola who is considered the father of mineralogy. In *De Re Metallica* ("On Metals"), first printed in 1556, Agricola describes the mining practices of his day in great detail, especially the techniques for exploiting and refining minerals and the procedures involved in the use of fusion to extract metals.

Modern crystallography, the study of the forms that compose crystals, was born between the second half of the 1600s and the end of the 1700s, thanks to the contributions of Nicholas Steno, Carangeot and Romé de l'Isle.

In 1801, Abbé René-Just Haüy discovered that minerals are composed of countless "molecules" that exactly reproduce the shape of their crystals, anticipating important discoveries that would only be confirmed a century later. During the 19th century, numerous scientists investigated the chemistry of minerals, among them Swedish chemist Jöns Jakob Berzelius, who established the principles of modern mineral classification.

The beginning of the 20th century marks a fundamental step in the history of mineralogy, with the discovery of the structure of

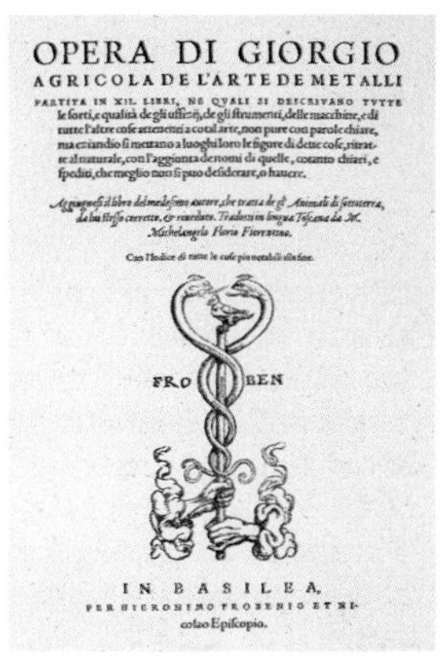

OPERA DI GIORGIO
AGRICOLA DE L'ARTE DE METALLI
PARTITA IN XII. LIBRI, NE QVALI SI DESCRIVANO TVTTE le sorti, e qualità de gli uffici, de gli strumenti, delle macchine, e di tutte l'altre cose attenenti a cotal arte, non pure con parole chiare, ma eziandio si mettano a luoghi loro le figure di dette cose, ritratte al naturale, con l'aggiunta de nomi di quelle, cotanto chiari, e spediti, che meglio non si puo desiderare, o hauere.

Aggiugnesi il libro del medesimo autore, che tratta de gl' Animali di sotterra, da lui stesso corretto, & riveduto. Tradotti in lingua Toscana da M. Michelangelo Florio Fiorentino.

Con l'Indice di tutte le cose piu notabili alla fine.

FRO BEN

IN BASILEA,
PER HIERONIMO FROBENIO ET NIcolao Episcopio.

Below left: *Frontispiece of* The Art of Metals *by Georgius Agricola.*
Below: *Portrait of the Abbé René-Just Haüy.*

minerals by German physicist Max von Laue in 1912. Experiments performed using X-rays proved for the first time that minerals are composed of atoms precisely arranged according to exact rules.

In the early 1960s, electronic microprobe analysis made possible the study of the chemical composition of minerals, and in comparatively short periods of time, enabled accurate chemical analysis of mineral fragments of even the smallest size (down to 4/100,000 of an inch or 0.001 mm). In the early 1970s, another highly advanced instrument, the transmission electron microscope, went into common use in laboratories and universities (at least those able to afford its high cost). Capable of magnifying millions of times, this tool opened a new frontier in mineralogy, allowing the direct observation of the atoms and the structures that compose minerals.

Below: *A beautiful crystal of eudialyte.*

THE FORMS OF CRYSTALS AND TWINNING

With few exceptions, all minerals are composed of atoms arranged following very precise rules. Where the natural conditions are favorable, these arrays of atoms develop smooth, flat surfaces and take on regular geometrical forms known as *crystals*. Crystallography is the study of these crystalline solids, including the principles that govern their growth, their exterior shapes and their atomic structures. Born as a branch of mineralogy for the study of the morphology of crystals, it has grown into a separate science that studies not only minerals but crystalline materials in general.

Simple home experiments—growing crystals of table salt by hanging a string in a dish of hot salted water and then watching crystals form on the string as

Above: *Wooden models of crystals with cubic symmetry.*
Right: *Halite crystals with visible signs of growth on the crystal faces.*

the water cools, or examining how ice crystals form in a freezer—reveal that solid substances can possess regular geometric forms, albeit forms that can vary greatly in their shapes. Anyone who has seen even a small collection of mineral specimens cannot help but notice a great variety of crystal shapes. In fact, the same mineral can even appear in different forms. The external appearance of a crystalline substance depends not only on its composition but also on the arrangement of its atoms and many other highly complex physicochemical factors, as well as the opportunity for the crystal to grow freely: in a rock composed of several different minerals such a granite, the mineral grains abut against one another and do not possess well-developed crystalline forms, instead assuming a more granular appearance.

Crystals are classified by the study of the forms identifying the arrangement of their faces. Since the end of the 1800s, several scientists (including Groth in 1895, Fedorov in 1925 and Rogers in 1935) have

Below left: *Quartz crystals that have freely developed inside a cavity.*
Below: *Specimen of granite from Baveno, Piedmont, Italy.*

Above: *A group of pyrite crystals showing a perfect example of cubic symmetry.*

used precise mathematical laws to establish 48 crystalline forms in minerals. These laws are related to the symmetry with which the structure of crystals is described. Take, for example, a crystal of pyrite with a cubic shape, and use it to perform two symmetrical operations: the first using an "ideal" axis to perform rotations and the second using "ideal" planes to create mirror images. Rotating the crystal in order to bring identical faces into identical positions one, two, three, four or six times is called axial symmetry. Cutting the crystal into two absolutely equal mirror images reveals plane symmetry. The structure and symmetry of crystals can become highly complex and specialized, but this example demonstrates how such operations make it possible to ascribe any crystalline form to a group of faces with very similar characteristics.

Right: *Crystals of stolzite that clearly illustrate the form of a tetragonal bipyramid.*
Far right: *Mimetite crystal, clearly showing which faces form a bipyramid, which form a hexagonal prism.*

The 48 known crystalline forms are divided into crystal systems. These are in turn divided into 32 classes, which are further subdivided into seven systems: triclinic, monoclinic, orthorhombic, tetragonal, trigonal, hexagonal and cubic (isometric). Each of these systems includes a certain number of crystal forms. For example, the monoclinic system includes the pinacoid, monoclinic prisms and a few other forms, while the cubic includes 16 forms, several of them with many faces. Among these is the hexoctahedron, composed of 48 faces.

Among the 48 crystalline forms, the most commonly observed in minerals include the pinacoid, formed by a pair of parallel faces and present in all the systems except the cubic; the prisms (composed of 3, 4, 6, 8 or 12 faces), also common to all the systems except the triclinic and cubic; and the pyramids (with 3, 4, 6, 8 or 12 faces) and bipyramids, (including 6, 8, 12, 16 or 24 faces), present only in the orthorhombic, tetragonal, trigonal and hexagonal systems. Among the less common forms is the rhombohedral, composed of six equal faces to form a distorted cube, found only in the trigonal system. The cubic system merits its own discussion since it features exclusive forms, of which the most common are the cube, the octahedron, the rhombododecahedron and the tetrahedron.

Above: *Typical crystal of hydroxylapatite with prismatic (hexagonal) faces.*
Above right: *Beautiful crystal of rhodochrosite with a rhombohedral habit.*

According to recent statistics, known minerals (4,000 mineralogical species are officially approved) are distributed in the crystal systems in the following percentages: 2 percent triclinic, 21 percent monoclinic, 20 percent orthorhombic, 12 percent tetragonal, 19 percent hexagonal and trigonal, and 26 percent cubic. As one can see, the greatest number of mineralogical species is concentrated in the cubic system, the one that possesses the greatest symmetry.

Twinned crystals are unique crystalline forms composed of the symmetrical growth of two or more crystals of the same mineralogical species. Such formations are not random, and the aggregation of two of more crystals in a twin follows certain crystallographic laws, known

as twin laws. These laws are defined by operations of symmetry. If the twinned crystals share a common face, they are called contact twins; if one crystal is piercing the surface of the other, they are called penetration twins; if several crystals are arranged along parallel planes in the same orientation they are called polysynthetic (or lamellar) twins; and several crystals arranged to simulate a single crystal and forming a complete circle are called cyclic twins.

Classical examples of minerals found as contact twinned crystals are gypsum, quartz and spinel. Penetrating twins are typical in fluorite, pyrite and orthoclase; polysynthetic twins are typical of albite; while cyclic twins can be found in aragonite, bournonite, cerussite, chrysoberyl, strontianite and witherite.

Top, left to right: *Crystals of bournonite twinned by penetration; crystals of cassiterite twinned by contact; fluorite crystal, a classical example of penetration twins with the rotation of two crystals.*
Above: *Crystal of quartz twinned by contact.*

PHYSICAL PROPERTIES

The physical properties of a mineral are a direct result of its chemical composition and atomic structure. Many of these physical properties can be determined easily, whether through observation or simple tests. Such properties, easily recognizable or easy to determine in the examples that follow, are very useful when trying to identify a mineralogical species. Of course, the written descriptions and accurate photographs presented in the mineralogical entries of this book will also be a great aid in identifying minerals. There is thus little reason to dwell on mineral properties that require the use of special devices, such as transmitted-light or reflected-light microscopes, or elaborate preparations of specimens for examination in order to make a diagnostic observation.

Below: *Boulangerite with characteristic thin acicular crystals, as flexible as hair.*
Right, top to bottom: *Rutile with a prismatic acicular habit; kyanite with a laminar crystal habit; and hübnerite with a typical tabular habit.*

Mineral habit

The habit of a mineral, meaning the predominant shape of its crystals as well as the way it forms crystalline aggregates, can be an excellent visual aid in identification. It often happens that the same mineralogical species will adopt different crystalline habits according to the conditions under which it forms.

In general, minerals form either distinct crystals or aggregates, and certain easily understood terms are now commonly used to describe the habits of these crystals and aggregates. The following adjectives are those most commonly applied to minerals found as distinct crystals:

Acicular (needle-shaped) This term is applied to isolated needle-shaped crystals, which are formed by such minerals as aragonite, artinite, aurichalcite, cyanotrichite, kermesite, malachite, plattnerite and scolecite.

Capillary or filiform These are hairlike crystals, thin and flexible, typical of such minerals as boulangerite, millerite and mesolite.

Laminar (bladed) Term applied to elongate, flat crystals, shaped like the blade of a knife; these crystals are typical in such minerals as actinolite, kyanite, covellite, molybdenite and tremolite.

Tabular A great number of minerals exhibit this habit, forming somewhat flat crystals but with obvious thickness and rectangular or square contours. The many minerals that form tabular crystals include anhydrite, autunite, barite, bertrandite, brookite, enargite, epidote, ferberite, heulandite, hübnerite, ferrocolumbite/manganocolumbite, pyrrhotite, rhodonite, stilbite and torbernite.

Different terms are applied to the aggregates formed by minerals. These include the following.

Dendritic, arborescent and coralline These terms refer to minerals that form branching, ramified aggregates, as in the case of silver, goethite and pyrolusite.

Fibrous and radiating These terms are applied to crystals joined in fibers, commonly forming radiating aggregates, as occurs with bavenite, cookeite, erythrite, hedenbergite, natrolite, okenite, pectolite, pyrophyllite, scolecite, thaumasite and wollastonite.

Globular, mammillary, botryoidal and spheroidal These similar terms are applied to aggregates composed of roundish crystals that rise from a common base. The different terms used to describe the shape of the crystals are: globular or mammillary (breast-like), botryoidal (like a bunch of grapes) and spheroidal (sphere-like). Such habits are typical of minerals like arsenic, brucite, hemimorphite, goethite, mimetite, pyrolusite, pyromorphite,

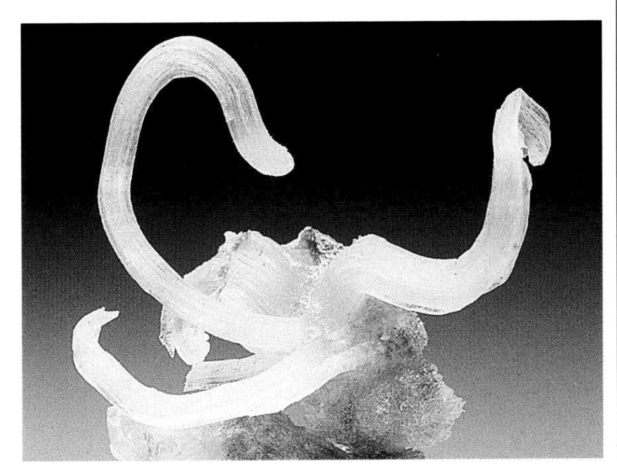

Top: *Arborescent aggregate of silver.*
Above left: *Mammillary aggregate, a characteristic well represented by malachite.*
Above: *Crystals of muscovite with a foliate, or micaceous, habit.*
Left: *Gypsum can form aggregates with a coralline habit.*

prehnite, rhodochrosite, talc, thomsonite, uraninite and wavellite.

Granular Term applied to a mineral without distinct crystalline form but which forms a mass with a typical granular appearance.

Lamellar, micaceous or foliate Flat, thin crystals, in some cases flexible and thin as a leaf, as exhibited by minerals in the mica and chlorite groups.

Massive This term is applied to compact minerals without any form or other distinctive characteristic.

Microcrystalline Many minerals form thin incrustations on rocks or on other minerals but are actually composed of aggregates of microcrystals, observable only if viewed with strong magnification (for example with a 20-power hand lens).

Skeletal or reticular Under certain conditions some minerals grow in aggregates of long, thin crystals resembling a net. Among these minerals, most of them native elements, are silver and bismuth, some sulfides, such as galena and skutterudite, and oxides such as cuprite.

Stalactitic and columnar These terms are applied to minerals with crystals that join to form cylindrical, columnar or conical aggregates, as exhibited by such minerals as aragonite, chalcanthite, calcite, descloizite, goethite, graphite, hydrozincite, opal, pyrolusite, rhodochrosite, halite and smithsonite.

Above: *Rhombohedral crystals of calcite with clear cleavage planes.*

Cleavage

Cleavage is a property that is easily visible, particularly on pieces of crystal that already have small breaks along the edges. Naturally, a cleavage test should not be performed on minerals forming perfect crystals; for observation purposes, it is always best to use already broken fragments. The term *cleavage* indicates the quality of a crystal to break along parallel planes, forming faces that are actually or potentially compatible with the symmetry of the crystal itself. This property is well developed in some minerals, such as the micas, in which the cleavage is absolutely perfect. Beryl exhibits indistinct cleavage, and quartz exhibits no cleavage at all. The directions in which cleavage occurs, or of breakage along a planar surface, depend directly on the structural arrangement of the mineral's atoms. The quality of cleavage is expressed by adjectives such as perfect, good, poor or indistinct; while the direction along which the cleavage occurs is defined according to the shape that results: cubic, octahedral, prismatic and so forth.

Where cleavage is absent—that is, where the surface that results on breaking a mineral is irregular—the term *fracture* is used. Fractures are described depending on the type of surface that results: conchoidal, in which the break takes place along a curved surface, such as in quartz or glass; splintery, for minerals like chrysotile; hackly, if the fracture produces a jagged surface, as in some minerals like silver and copper; and uneven, if the surface is irregular, as in descloizite, garnets and tourmaline.

Above left: *Cleavage is absent in quartz, which has a conchoidal type of fracture similar to that in glass.*
Above: *Galena offers an example of perfect cleavage along parallel planes.*

Hardness

Hardness is a measure of a mineral's resistance to being scratched: it indicates the ease or difficulty with which a mineral can scratch or be scratched. In 1824, the German mineralogist Friedrich Mohs took ten common minerals and put them in order according to hardness, running from the softest to the hardest. Known as the Mohs scale (see bottom of page), it is still commonly used in mineralogy.

Mineral	Hardness
Talc	1
Gypsum	2
Calcite	3
Fluorite	4
Apatite	5
Orthoclase	6
Quartz	7
Topaz	8
Corundum	9
Diamond	10

Left: *Talc is the softest mineral on the Mohs scale of hardness (far left).*

Above: *Quartz, shown here in perfect transparent crystals, has a value of 7 on the Mohs scale.*
Above right: *Diamond is the mineral most resistant to scratching, with a value of 10 on the Mohs scale.*

The sequential order of this scale means that each mineral can scratch the ones before it (those with lower numbers) and be scratched by those following it (with higher numbers). Thus diamond can scratch all the others, while talc cannot scratch any. Again, it is best not to perform a scratch test on well-crystallized minerals, since doing so might ruin them; fragments of minerals, if possible already broken, are always better. Furthermore, if the test must be performed on small fragments of a mineral, it is always best to repeat the test several times and perform it with minerals that are both softer and harder, checking each time with a magnifying glass to see if there is any sign of scratching.

Specific gravity

A mineral's specific gravity or relative density (although the terms are not really synonymous since the latter involves a different unit of measurement) is simply a measurement that expresses the relationship between the weight of the mineral and the weight of a similar volume of water at the temperature of 39.2°F (4°C). This definition is at the root of Archimedes' Principle, and it means that a mineral with a density equal to 2 weighs twice as much as the equivalent volume of water. Determining specific gravity can be of great help in identifying a mineral, particularly when dealing with minute crystals or gems of an unknown nature.

A very useful tool for this purpose is the pycnometer, a small glass flask fitted with a stopper. The measurement is made first by weighing the flask empty (P), after which the fragment of mineral to be weighed is put in the empty flask and the flask is weighed again (M). The flask is then filled with distilled water and is heated a few minutes to remove any air bubbles. The flask with the mineral is filled to the top of the opening and weighed again (S). Finally the flask is completely emptied then filled with distilled water only and weighed (W). Using these four measurements the simple calculation illustrated at the top of the next page can be made. The result will be the specific gravity of the mineral.

Below: *The pycnometer is a useful and practical tool for measuring the specific gravity of minerals.*

$$\text{Specific gravity} = \frac{M - P}{W + (M - P) - S}$$

remembering that:
M = weight of empty flask + weight of mineral
P = weight of empty flask
W = weight of flask full of distilled water
S = weight of flask filled to the top with distilled water + weight of the mineral

Luster

The term *luster* refers to the appearance of a mineral's surface (for example, the face of a crystal) when illuminated with reflected light. In general, the types of observable luster are metallic and non-metallic. A mineral that exhibits a brilliant surface similar to that of a metal shows metallic luster and, in general, is opaque to light. A great many examples of this luster can be found among the native elements and the sulfides, such as silver, gold, chalcopyrite, galena, pyrite and pyrrhotite. Minerals that present non-metallic luster are often colored and translucent, meaning light passes through them, but they are not transparent.

Several terms are used to describe non-metallic luster:

Vitreous This term refers to a luster similar to that of glass; typical examples are beryl, quartz,

Above left: *Pyrite, like many other sulfides, has a metallic luster.*
Above: *Prismatic crystals of elbaite with a typical vitreous luster.*
Left: *Sphalerite with its typical resinous-adamantine luster.*

Above: *Diamond has an adamantine luster.*
Above right: *The luster of heulandite, a mineral of the zeolite group, can be defined as pearly.*

topaz and minerals from the tourmaline group.

Resinous This term is applied to a luster like that of resin; classical examples are sphalerite and sulfur.

Pearly This term applies to a luster similar to the iridescence of a pearl; it is characteristic of minerals that exhibit clear planes of cleavage, such as brucite, heulandite, the minerals in the chlorite group and the micas.

Silky This luster is characteristic of minerals with a silky appearance; typical examples are okenite, pectolite, thaumasite and wollastonite.

Adamantine This is a very brilliant luster, similar to that of a diamond; it is characteristic of minerals with a high index of refraction (an optical property); typical examples are anglesite, cassiterite, cerussite, phosgenite, linarite and scheelite.

Other properties

Minerals possess many other physical properties. Among these is fluorescence, the emission of visible light by a mineral when it is exposed to a luminous source of a particular wavelength. This experiment can easily be performed using normal ultraviolet lamps, with wavelengths of 365 nm (called UV long) or 254 nm (called UV short), being careful to never look directly at the light source. When placed under such light sources, some minerals emit visible light in the form of vivid colors. Fluorescence is a highly unpredictable property; specimens belonging to the same species as a fluorescent mineral but from different localities may exhibit no trace of fluorescence. Classical fluorescent minerals are autunite, fluorite (this strongly fluorescent mineral gives the property its name), calcite, diamond, opal and halite. Fluorescence is caused by the presence of certain metallic impurities like the rare earth elements at trace levels within the atomic structure of a mineral; these absorb part of the ultraviolet light and in turn emit luminous energy that our eyes see in the form of vibrant colors.

Some minerals behave as magnets and are attracted by instruments able to develop a magnetic field; magnetite is an example of such a mineral. This property can be easily observed by simply

Below, left to right: *Specimen of autunite photographed under natural light and the same specimen photographed under ultraviolet light.*
Bottom, left to right: *Specimen of aragonite photographed under natural light and the same specimen photographed under ultraviolet light.*

putting a magnet near the mineral to be tested; if magnetic, the magnet will immediately attach itself to the mineral. In reality, this property is found in many minerals, but in general can be detected only with the use of a strong electromagnet. Using an electromagnet, it is possible to separate out of sand the part composed of tourmaline, which is paramagnetic (susceptible to magnetism), leaving behind the part composed of quartz, which is diamagnetic (not susceptible to magnetism).

Optical properties are used a great deal in the study of minerals, but can be difficult to measure without special instruments. One of these properties, however, can be observed using a small amount of sufficiently transparent calcite crystal (the ideal would be a fragment of Iceland spar). The crystal is positioned above writing on a piece of paper and is turned; an observer viewing through the crystal will see two images of the writing. This represents the optical property known as double refraction: the light entering the crystal is split into two beams that then exit the crystal at different angles of refraction. Also known as rotary polarization, this aspect of crystals has important applications in the production of many kinds of optical equipment, including the special lenses for polarized-light microscopes.

Left: *Double refraction, the splitting of light across a crystal, is an optical characteristic highly visible in this clear cleavage rhomb of calcite (so-called Iceland spar), which doubles the image of the black line on the wall behind it.*

HOW MINERALS ARE STUDIED

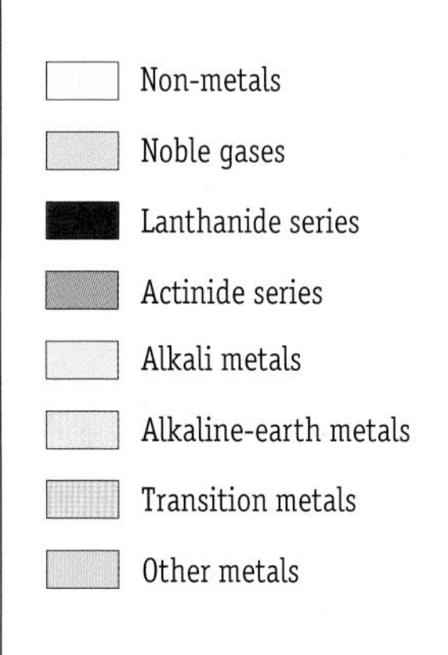

The atom and the periodic table

Knowing a mineral's chemical composition is of fundamental importance because the physical properties we have learned to recognize result in large measure on its chemical composition. The first step in understanding chemical composition is to understand the atom, the basic unit of matter and the smallest unit of a chemical element to have the properties of that element. The atom is composed of several components, including a nucleus composed of protons and neutrons and, orbiting the nucleus, electrons. Every proton has a positive charge (+), every electron has a negative charge (–), while every neutron, as indicated by its name, is electrostatically neutral. Every atom must be neutral, thus the number of protons must always be matched with an equal number of electrons. The simplest atom is hydrogen, composed of one proton and one electron. The atomic number of the atoms that make up a chemical element is the number of protons contained in its nucleus (which is matched by an equal number of electrons). The periodic table is composed of 92 natural chemical elements and a certain

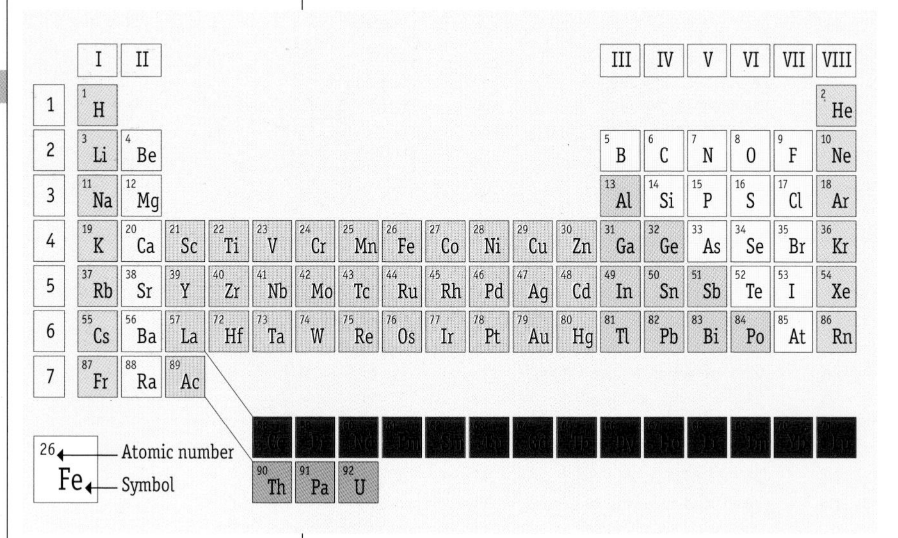

Above: *The periodic table of elements gives the symbol of each chemical element and its atomic number.*

number of artificial elements, arranged according to a series of atomic and chemical characteristics, including each element's atomic number (represented by the symbol Z). Each element has a chemical symbol, and all the elements are listed in order according to atomic number. Thus hydrogen, symbol H, which possesses one proton (Z = 1), is at the top left of the table; iron, symbol Fe, has 26 protons (Z = 26) and is

CHEMICAL ELEMENTS AND SYMBOLS

This list only includes the first 92 elements

Symbol	Atomic number (Z)	Name	Symbol	Atomic number (Z)	Name	Symbol	Atomic number (Z)	Name
Ac	89	Actinium	Ho	67	Holmium	Rn	86	Radon
Al	13	Aluminum	H	1	Hydrogen	Re	75	Rhenium
Sb	51	Antimony	In	49	Indium	Rh	45	Rhodium
Ar	18	Argon	I	53	Iodine	Rb	37	Rubidium
As	33	Arsenic	Ir	77	Iridium	Ru	44	Ruthenium
At	85	Astatine	Fe	26	Iron	Sm	62	Samarium
Ba	56	Barium	Kr	36	Krypton	Sc	21	Scandium
Be	4	Beryllium	La	57	Lanthanum	Se	34	Selenium
Bi	83	Bismuth	Pb	82	Lead	Si	14	Silicon
B	5	Boron	Li	3	Lithium	Ag	47	Silver
Br	35	Bromine	Lu	71	Lutetium	Na	11	Sodium
Cd	48	Cadmium	Mg	12	Magnesium	Sr	38	Strontium
Ca	20	Calcium	Mn	25	Manganese	S	16	Sulfur
C	6	Carbon	Hg	80	Mercury	Ta	73	Tantalum
Ce	58	Cerium	Mo	42	Molybdenum	Tc	43	Technetium
Cs	55	Cesium	Nd	60	Neodymium	Te	52	Tellurium
Cl	17	Chlorine	Ne	10	Neon	Tb	65	Terbium
Cr	24	Chromium	Ni	28	Nickel	Tl	81	Thallium
Co	27	Cobalt	Nb	41	Niobium	Th	90	Thorium
Cu	29	Copper	N	7	Nitrogen	Tm	69	Thulium
Dy	66	Dysprosium	Os	76	Osmium	Sn	50	Tin
Er	68	Erbium	O	8	Oxygen	Ti	22	Titanium
Eu	63	Europium	Pd	46	Palladium	W	74	Tungsten
F	9	Fluorine	P	15	Phosphorus	U	92	Uranium
Fr	87	Francium	Pt	78	Platinum	V	23	Vanadium
Gd	64	Gadolinium	Po	84	Polonium	Xe	54	Xenon
Ga	31	Gallium	K	19	Potassium	Yb	70	Ytterbium
Ge	32	Germanium	Pr	59	Praseodymium	Y	39	Yttrium
Au	79	Gold	Pm	61	Promethium	Zn	30	Zinc
Hf	72	Hafnium	Pa	91	Protactinium	Zr	40	Zirconium
He	2	Helium	Ra	88	Radium			

located near the center; and uranium, symbol U, which has 92 protons (Z = 92), comes last, at the end of the table.

The organization of the atomic structure, and therefore of the chemical elements that compose a mineral, depend on the way in which the atoms combine with other atoms. These combinations are made possible through a series of chemical bonds, the principal types of which are

Above: *Alphabetical list of the chemical elements with their symbols and atomic numbers.*

ionic, covalent and metallic. These bonds are electrical forces, and leaving aside the complex chemical explanations by which they are identified, it can be stated that many of the bonds in minerals are ionic or metallic. Minerals resulting from ionic bonds have a moderate hardness and density, while those resulting from metallic bonds, as the term suggests, exhibit metallic characteristics and are high in density and low in hardness, aspects typical of such minerals as silver, gold, copper and many sulfides. Purely covalent bonds are somewhat rare but are the strongest. Diamond is an example of a mineral that results from this type of bond; it exhibits very great hardness and is able to scratch any other mineral.

Modern methods of analysis

Until the early 1950s the study of the chemistry of minerals was performed primarily using so-called "wet-test" methods of chemical analysis, meaning the highly laborious dissolution of minerals in acids. The elements present were determined using a series of titrations and chemical reactions using specific reagents.

The 1960s saw the beginning of great progress in scientific knowledge, thanks in part to the introduction of new laboratory instruments. The scanning electron microscope, able to magnify crystals many thousands of times, soon became widespread. In combination with the electron microprobe, it made chemical analysis possible on a microscopic scale, without destroying solid inorganic substances such as minerals. Such microanalyzers offer great advantages since they enormously shorten the time necessary for chemical analysis and make possible accurate analyses of fragments of minerals, even those only a few micrometers in size. Most of all, however, they allow us to monitor and measure even tiny changes in the chemical composition of a crystal, a procedure that was absolutely impossible under the old wet-test methods of analysis.

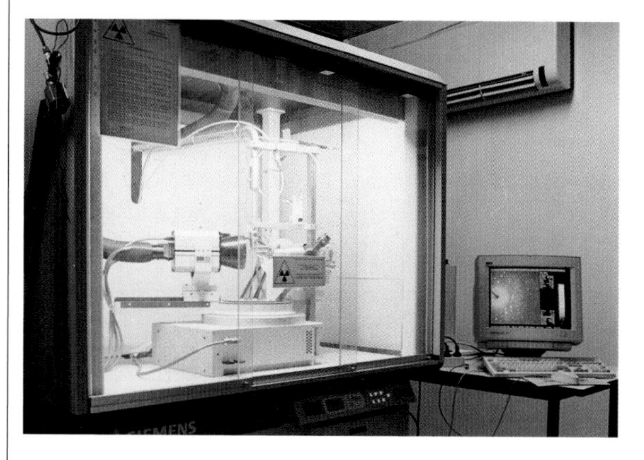

Right: *The X-ray diffractometer is an indispensable instrument for studying the structure of minerals (courtesy of the Department of Molecular Chemistry and Inorganic Stereochemistry of the University of Milan, Italy).*
Below: *Diamond is one of the few minerals to possess a purely covalent bond.*

X-rays, discovered by German physicist Wilhelm Conrad Roentgen in 1895, have made a vital contribution to the study of minerals; in 1912, Max von Laue applied the radiation produced by X-rays to the study of crystals. Until then, crystallographers had used cleavage, various optical properties, and most of all the meticulous study of the morphology of crystals in order to determine the organization of the structure of minerals. With those tools they had been able to determine the structure of many minerals and had done so with a certain degree of accuracy. Yet they lacked a means by which to prove that the minerals possessed a symmetrical and periodic (or regularly recurring) internal organization. The application of X-rays finally made it possible to measure the distances between atoms and to locate their positions inside a crystal, thus obtaining extremely important experimental data.

The laboratory instrument that determines and records the positions of a mineral's atoms is the X-ray diffractometer, an instrument capable of collecting and recording the crystallographic data of a mineral using both tiny crystalline fragments and minute portions reduced to a powder.

Over the past 20 years, these modern instruments have enabled scientists to increase the number of mineralogical species identified from

2,000 to over 4,000. This number will no doubt continue to grow, since about 40 new minerals are discovered and studied each year. Because of their high cost, such instruments are usually found only in research institutes, universities and museums throughout the world. If you happen to come across a strange mineral, you should try enlisting their help—you may find yourself taking part in the discovery of a new species.

Left: *To the right in the photo is a scanning electron microscope (SEM) which, combined with energy-dispersion spectroscopy (EDS), makes possible the precise study of the chemical composition of minerals (instrument in use at the Museum of Natural History, Milan).*
Above: *Image of a group of crystals of hingganite-(Y) obtained using an electron microscope.*

THE ENVIRONMENTS IN WHICH MINERALS FORM

The natural environments in which minerals form are numerous and have markedly different characteristics of temperature, pressure and chemical composition. Some minerals, such as quartz or calcite, can form in many different environments, including surface water, deep hydrothermal veins, and as intrusions in magma; others, such as kyanite or glaucophane, form only in specific environments over specific ranges of pressures and temperatures.

If one were to descend from the surface of the earth toward its center, one would encounter rocks of increasingly higher temperatures exposed to increasingly elevated levels of pressure to finally reach the bottom of the so-called continental crust, which has a depth of between 12 and 25 miles (20 and 70 km). Beyond that depth, there are other types of rocks, formed in temperatures of over 1,800°F (1,000°C) and under even higher pressures; these rocks are part of the earth's upper mantle.

As geological studies have shown, the rocks, and thus minerals, that are found on the surface of the earth were formed at different times and under widely different ranges of pressures and temperatures. This phenomenon is explained by the enormous geodynamic processes behind plate tectonics. Over millions of years, plate tectonics have profoundly changed the terrestrial surface, causing the subsidence (subduction) of large sections of oceanic crust (the earth's crust on ocean bottoms), the thickening and deformation, through metamorphism, of large sections of continental crust, the formation of chains of volcanoes (volcanic arcs), and the creation of submarine basins with the formation of new oceanic crust. On the surface, these phenomena have caused the formation of mountain chains, rift valleys and sedimentary basins of various sizes.

The processes of erosion play a fundamental role in exposing rocks and minerals formed below on the earth's surface. In fact, the faster a mountain chain rises as a result of the collision of two areas of continental crust, the faster the erosion, with consequent exposure on the surface of rocks formed at great depths. Erosion is also responsible for the refilling of rifts and submarine basins with thick layers of sediment.

As a result of geodynamic processes taking place at varying depths of the crust, rocks and minerals are in continuous formation and transformation, undergoing erosion and eventual destruction on the surface. The same mineralogical species that formed at the dawn of our planet's history, more than four billion years ago, can still form today and will continue to form so long as the geodynamic and erosional processes are active. However, some geological environments, and therefore also certain situations of pressure, temperature and composition, belong only to epochs of the past and are therefore unique and unrepeatable. A good example of this is komatiitic lava, which contains a variety of olivine known as "spinifex" formed more than two-and-a-half billion years ago,

Below: *Metamorphic rocks formed at a depth of about 25 miles (40 km) and at temperatures of around 1,500°F (800°C) in an outcrop in southern Madagascar.*
Bottom: *Crystal of calcite, a widespread mineral on our planet.*

a time when the earth's internal temperature was greater than it is today and its continental crust was far thinner.

Above left: *Quartz is a widespread component of many igneous and metamorphic rocks.*
Above: *Corundum crystal contained in gneiss formed in a metamorphic environment at high temperature and pressure within the earth's crust, here exposed in a locality in southern Madagascar.*

The formation of mountain chains and the formation of minerals

The formation of minerals is a vast subject, and this introduction will limit itself to a brief overview—general information to give the reader an idea of the primary environments in which the magnificent crystals photographed in this book were formed. The following examples are framed within the evolutionary processes of a mountain chain, including the magmatic, metamorphic and sedimentary processes associated with hydrothermal activity and the phenomena of erosion and surface alteration.

Magmatic environment

The entire evolution of a mountain chain—through its birth to the main period of its growth to the period of its subsidence and collapse—is accompanied by the production of magma, which rises from the depths of the earth's crust and upper mantle and, cooling, generates igneous ("fire-formed") rocks.

Not all magma is thrown onto the earth's surface through volcanic activity. The magma that remains within the crust can accumulate and cool, forming intrusive rocks. Depending on their size and shape, intrusive rocks are known as sills, dikes, stock, plutons and batholiths; the latter being the largest. A reservoir of magma within the earth's crust is called a magma chamber. The cooling of

Below: *Granite pluton outcrop in southern Madagascar.*

Right: *Fumarole in the Phlegrean Fields near Naples, Italy.*
Below: *Sulfur and other newly formed minerals produced by gases from volcanic craters.*

magma in a chamber can result in a pluton or a batholith.

Magma that rises through the crust and is erupted onto the surface by volcanic activity forms extrusive rock; lava that flows onto the surface without explosive activity forms effusive rock. Igneous rocks in veins or dikes usually result from modest-sized masses of magma that have cooled very near the surface without being erupted.

The magmas produced during the evolutionary phases of a mountain chain assume a certain geochemical "signature," generating so-called magmatic series (associations of typical magmatic rocks). The magmas produced during different phases in the development of the chain will have different chemical compositions, with varying amounts of aluminum, silicon, sodium, potassium and calcium. With this data, geologists studying the chemical makeup of igneous rock can determine whether the rock was formed during an initial phase in the construction of the mountain chain, during the phase when the chain was at the height of its growth, or during a later phase of subsidence. As magma rises through the crust and then as it cools in a magma chamber, it goes through a complex process of crystallization, and because of the variations from place to place, different magmas produce different rocks. The most primitive magma generally have a more basic composition and, once crystallized, form rocks with a somewhat dark coloration with an abundance of such elements as iron, magnesium and calcium, to form dark minerals like biotite and the amphiboles and pyroxenes. More evolved magmas are generally more acidic and, once crystallized, form rocks with a pale coloration high in such elements as silicon, aluminum, potassium and sodium, to form pale minerals like quartz, potassium-rich feldspars and plagioclase feldspars. Typical associations of intrusive rock, from the most basic to the most acidic, include gabbros, diorites, granodiorites, granites, aplites and pegmatites. Other intrusive associations, typically rich in alkaline elements (sodium and potassium) in the evolved members, are gabbros, monzonites, granites, syenite and nepheline syenites. Common effusive rocks, variously associated, include basalts, andesites, phonolites, dacites and rhyolites.

During the crystallization of magmas, favorable environments, usually rich in aqueous fluids and at high temperatures, lead to minerals that grow large-size crystals instead of simply crystallizing as the components of rock. Magma clearly creates an enormous variety of environments for mineralogical formation.

Above left: *Fluorapophyllite and stilbite-Ca, minerals found in the cavities of basalts in India.*
Above: *Beryl and albite, beautiful examples of crystals found in granitic pegmatites in Pakistan.*

Pegmatites are among the magmatic environments most favorable to the formation of magnificent crystals. Pegmatites, in fact, are coarse-grained rocks composed principally of quartz and feldspar that may have cavities left by fluids that concentrated in the core of the vein. Within these cavities, the crystals of many rare minerals, including some of interest as gemstones (tourmaline, beryl, topaz), have the opportunity to develop undisturbed into spectacular forms.

Below: *Alluvial detritus deposits from the Permian period (about 290 million years ago) exposed in the south of Madagascar, a result of the erosion of an ancient mountain chain.*

Sedimentary environments

Erosion is responsible for the subsidence of mountain chains. The products of the alteration of rocks and fragments of minerals (clasts) are transported by surface water, ice and wind to form alluvial deposits of detritus on the outer borders of the mountain chain. Transported to even more distant zones, these materials accumulate in layers within submarine basins and tectonic depressions (particularly in areas where the continental or oceanic crust is undergoing subduction caused by plate tectonics), where they can form large-scale sedimentary deposits.

The dissolved minerals, clays and clasts that settle onto the sea floor are joined by a great abundance of many chemical elements, and these, under suitable conditions, crystallize, creating deposits through chemical precipitation. An important contribution to the formation of sedimentary deposits is made by fragments of the shells of microorganisms, which form their own bioclastic sediments.

Over the course of millions of years, the accumulated sediment undergoes a series of physical and chemical changes known as diagenesis, which involves pressure exerted by the upper layers on the lower, and an increase in temperature caused by the heat rising from within the crust. The result of diagenesis is that the variously deposited loose sediments cohere and solidify to form sedimentary rocks.

Above: *Exceptional crystal of transparent gypsum from evaporite deposits in the Apennines mountains of Italy.*

Sediments that form through chemical precipitation in restricted saltwater basins or lacustrine environments are called evaporite deposits. During diagenesis, the circulation of water through evaporite deposits can cause the formation of cavities, in some cases large in size, and spectacular crystals can form within these cavities, notably gypsum, halite, aragonite and sulfur. Examples of evaporite deposits of this type have been found in the Apennine Mountains of Italy and in Sicily, dating to the Messinian age. During the Messinian (around six million years ago), the Mediterranean Sea was cut off from the Atlantic and underwent an intense process of evaporation, leaving thick evaporite deposits in sedimentary rocks throughout the area.

As a result of the volcanic activity typically associated with the processes of the formation of mountain chains, layers of volcanic ash can become deposited among the layers of sediment, and these intercalations can later form important metalliferous deposits. Through the process of diagenesis, the metal ions in the ash, such as zinc, lead, copper, iron, arsenic, antimony, vanadium and chromium, as well as fluorine and barium, combine with sulfur and oxygen to form a large variety

Left: *This cluster of acicular crystals of aurichalcite formed inside metalliferous veins of the Triassic age in an outcrop in the pre-Alps of Lombardy, Italy.*

of minerals (sulfides, sulfates, fluorides, vanadates, chromates) and to form polymetallic deposits. This environment favors the formation of cavities with magnificent crystals of minerals of the so-called gangue, such as fluorite, barite and calcite, associated with metallic minerals like pyrite, chalcopyrite, sphalerite and arsenopyrite. When erosion exposes deposits of such primary minerals, they tend to alter through interaction with groundwater. The result is the formation of an oxidized zone and splendid new crystals of new minerals, including malachite, azurite, anglesite and cerussite, may then be found in the cavities of such rock.

The karst phenomenon occurs when, during the erosion of carbonate sedimentary rocks such as limestone and dolomite, groundwater dissolves the rocks to create unusual formations, including underground streams, ravines, spires and sink-holes known as dolines.

Over time, the underground circulation of water that results from karst phenomena forms complex systems of caverns. The calcium carbonate is redeposited by water in these settings to form fantastic stalactites and stalagmites. In certain rare cases, as in certain sites in Mexico, gypsum also forms in the grottoes, in some cases producing transparent crystals of enormous size.

Below: *This panorama of mountain crests provides an example of contact between a pluton of granitic composition (to the left) with sedimentary rocks (to the right), which have been transformed by the heat of metamorphic skarn.*

Metamorphic environment

The large mountain chains on our planet were formed through the collision of areas of crust driven against each other by the forces of plate tectonics. All the rocks that constitute the crust and that are involved in the formation of a mountain chain undergo high pressure and temperature as well as chemical activity. In this way the minerals that constitute the rocks, whatever their nature, are

exposed to pressures and temperatures much different from those under which they formed. The result of these intense forces is that the minerals reform as species that are stable in the new environmental conditions. These processes of transformation are called metamorphism. The presence of fluids, usually aqueous or rich in carbon dioxide, is necessary for these mineralogical reactions to take place; the fluid provides the element mobility, the means for the chemical elements to migrate from the crystal of a mineral that is coming apart to the crystal of a newly-formed mineral.

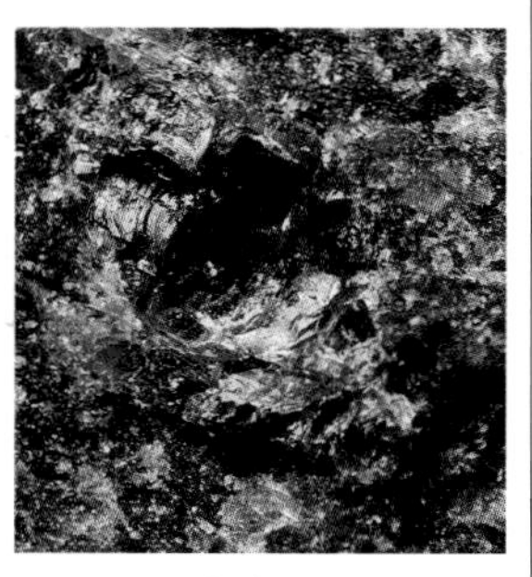

Top: *Rhombododecahedral crystals of grossular formed through contact metamorphism.*
Above: *Inclusion of red crystals of pyrope also formed through metamorphism.*

During the processes of convergence between the margin of a plate of oceanic crust, comparatively thin and pliable, and the margin of a plate of thicker and more rigid continental crust, the oceanic plate tends to be subducted beneath the continental plate. An "accretionary wedge" is formed between the two plates. This is a jumbled mass of rocks composed of buoyant material scraped off the oceanic crust that remains at the surface and sedimentary material accumulated in abundance due to the rapid disaggregation of the highest parts of the mountain chain being constructed. This accretionary wedge, located at the point of greatest pressure where the two crustal plates meet, is dragged partially downward and tends to undergo intense deformation. If the convergence of the plates goes on for a long time (for example, tens of millions of years), it is possible that the oceanic plate will be entirely consumed and that a new margin of continental crust will become juxtaposed with the one under which the oceanic crust was subducted. This new collision will bring an end to the process of convergence since neither of the two continental plates can subduct under the other owing to the buoyancy of each. Crushed between the two margins of crust, the accretionary wedge will be driven upward. The rocks and minerals involved in these processes can be brought under enormous pressures at depths of more than 60 miles (100 km) and at temperatures that exceed 1,800°F (1,000° C). There are thus intense phenomena of mineralogical transformation and in some cases fusion associated with the formation of magmas that rise toward the earth's surface.

A geological situation as complex as this generates many different environments for the formation of minerals, leading to an exceptional variety. Once again, the presence of abundant fluids, most of all water, is of importance for the formation of well-formed, large crystals. Examples of exceptional crystals formed in metamorphic environments are numerous in the Alpine chain. There are, for example, the large, perfect crystals of almandine garnet, contained within a schistose rock rich in mica, from the Passo del Rombo in the Trentino-Alto Adige region of Italy, or the splendid crystals of azure-blue kyanite with red-brown staurolite, contained in a pale, fine-grained micaceous rock from the summit of Pizzo Forno in the Ticino Canton of Switzerland.

A particular type of metamorphism, known as contact or thermal metamorphism, occurs when a large mass of magma is intruded into the earth's cool upper crust. The heat released by the magma while cooling heats the surrounding rocks, causing mineralogical reactions and the formation of new crystals of characteristic minerals, such as grossular, epidote, diopside, scapolite and vesuvianite.

Hydrothermal activity

Within the environments in which minerals form, the growth of well-formed crystals is always favored by the presence of fluids, most commonly aqueous fluids. The circulation of water at a more or less high temperature within any type of rock is known as hydrothermal activity.

Hydrothermal activity occurs wherever a source that modifies the normal distribution of heat is present at some depth of the earth's crust. Such heat often results from the intrusion of a mass of magma that has risen from below.

Hydrothermal fluids can move through rock of magmatic, sedimentary or metamorphic origin, and as they do, transport many chemical elements and deposit them in veins and fissures. This type of environment is ideal for the formation of exceptional crystals of a great many mineralogical species. According to the chemical composition of the fluids, a different mineral association can be formed wherever there is enough space. For example, there can be veins of quartz, barite, fluorite, of metallic minerals, or alpine-type veins. The last-named, of particular interest to mineral collectors, usually form within fractures in metamorphic rocks, forming typical crystals not only of quartz and hematite in the iron-rose-crystal form, but also titanite, epidote, albite, anatase, rutile and apatite.

Above: *Titanite is a mineral typical of Alpine hydrothermal veins.*

The sulfur mines of Sicily

There are large deposits of sulfur on the island of Sicily, most of all in the provinces of Agrigento, Caltanissetta and Enna. These were the island's major economic resource during much of the 19th and 20th centuries. The deposits of Sicilian sulfur belong to a sedimentary formation known as the Sulfur-bearing Gypsum Series, which extends more or less unbroken from north to south along the entire chain of the Apennine Mountains. This formation, from the middle to upper Messinian age (about six million years old), was formed in evaporite environments through the effect of a strong evaporation of seawater following the closing of the Strait of Gibraltar, isolating the Mediterranean basin from the Atlantic Ocean. The artisan methods used to extract the sulfur made possible the recovery of many exceptional mineralogical specimens, most of all sulfur. Certain mines in particular—such as Cozzo Disi, Floristella, Giumentaro Capodarso and Gibellina—have become world-famous for producing spectacular examples of crystals. The sulfur crystals from the deposits of Sicily are among the best examples ever found of this mineral species. The crystal forms, constituted by rhombic bipyramids, rhombic prisms and pinacoids, form beautiful, perfect crystals with a bipyramidal habit, usually terminating in pinacoid faces. There are also crystals with a tabular habit or with rounded edges, the latter characteristic being common in larger-size crystals. The most interesting minerals associated with sulfur include aragonite and celestine. The mineralogical examples of these two species are also considered among the best in the world, and along with the sulfur they enrich the principal museum collections of the world. The crystals of aragonite appear as perfect prisms with hexagonal contours. Close study, notably including observation of the striations on the faces of the basal pinacoid of these crystals, has revealed that they are composed of the cyclic twinning of three crystals with rhombic symmetry rotated at 60° from one another.

The Floristella mine has produced the best specimens of celestine, with groups of very elongated prismatic crystals, white or colorless, that form beautiful specimens on a matrix rich in sulfur.

Several decades have now passed since these mines ceased activity, making it highly doubtful that examples equal to those already found will ever come to light.

Top: *Group of crystals of white celestine, associated with sulfur, from Cianciana, Sicily.*
Above: *Splendid specimen of sulfur crystals from Sicilian mines.*

Pegmatic minerals from Baveno, Italy

As long ago as 1500, the area of Baveno in the northern Italian region of Piedmont, was famous throughout Europe for the extraction of pink granite for ornamental use. The area is also the site of spectacular cavities of orthoclase, quartz, albite and fluorite, often associated with a large number of other minerals, some of which were first found in that locality, such as bavenite, bazzite, calcio-ancylite-(Nd), cascandite, jervisite and scandiobabingtonite.

Below: *Perfect octahedral crystal of fluorite from granite of Baveno, Italy.*
Center: *Twinned crystal of orthoclase from Baveno.*
Bottom: *View of active granite quarry near Baveno.*

The orthoclase crystals from Baveno, certainly among the most important specimens from this area, are among the most stunning ever found of this mineral.

Orthoclase forms pink or white crystals with a prismatic habit that are of various lengths; they are commonly twinned following the Baveno Law or more complex laws of twinning. In Baveno, orthoclase is associated with all the other late minerals that are formed in the miarolitic cavities of granitic pegmatites; fluids rich in metallic elements left over from the crystallization of the granitic magmas concentrate in these cavities. Even today it is possible to find good mineral specimens in this area, although the closing of many of the local quarries during recent decades has done much to diminish opportunities. The still-active quarries include one in the small town of Oltrefiume, which has furnished excellent specimens since the 18th century, including the first known examples of orthoclase, which in 1779 were fully described and illustrated in his *Memoirs* by Ermenegildo Pini, a Barnabite friar and Professor of Natural History in Milan.

Although the mineralogical wealth of Baveno has been well known for a long time, new and rare species are continually being discovered, as a result of the assiduous research carried out by collectors and the analyses conducted in recent years by important universities and scientific institutions. These discoveries continue to enrich the knowledge and scientific interest in this famous mineralogical locality.

Top: *Panorama of iron mines near Rio Marina on the island of Elba, Italy.*
Above: *Exposed pegmatite vein on Mount Capanne on Elba.*
Below left: *Beautiful crystal of elbaite from Elban pegmatite.*
Below right: *Crystal of pyrite from the mines of Rio Marina.*

Minerals from the island of Elba, Italy

Because of its ancient mineralogical tradition, the island of Elba is looked upon as a kind of El Dorado, both by collectors from all over the world—for the presence of numerous minerals characterized by exceptional crystallization—and by many scholars, who have identified and described more than 170 mineralogical species from this island.

The particular geological conformation of Elba, the result of intense recent magmatic activity of an intrusive character, has permitted the formation of important metalliferous deposits, particularly associated with skarn. The exploitation of these mineral resources, above all the extraction of iron, began in ancient times, as indicated by literary citations, notably from Aristotle, Virgil and Strabo, who make ample references to the local population's exploitation of the island's iron deposits. Today the iron mines are all inactive, but those of Rio Marina, Terra Nera and Capo Calamitta have produced spectacular specimens of hematite, ilvaite and pyrite that enrich the holdings of the most important mineralogical museums throughout the world.

Along the western side of the intrusive monzogranite massif of Mount Capanne appear numerous pegmatite veins famous throughout the world for magnificent specimens of elbaite, beryl and orthoclase, associated with numerous other accessory minerals. Most of the tourmaline veins on the island that became famous, such as Grotta d'Oggi, La Speranza and Fonte del Prete, were exploited before World War I. Even so, during recent decades, the careful efforts of various scholars and private collectors has made possible the discovery of new veins that have provided specimens comparable to those of the past.

THE CLASSIFICATION OF MINERALS

The classification of minerals is in a continual state of evolution. Every year new minerals are discovered, others are redefined as the result of further study, while others are discredited. The 288 minerals described in this book are only a small portion of the 4,000 or so known species.

Those that have been included are all officially recognized by the Commission on New Minerals and Mineral Names (CNMMN) of the International Mineralogical Association (IMA), a commission composed of mineralogists from around the world that oversees the introduction and the nomenclature of new minerals. Among other goals, the commission seeks to avoid the proliferation of poorly defined or inadequately studied species. Its official website is www.geo.vu.nl/users/ima-cnmmn

Below: *Gold nugget from Australia weighing 5½ ounces (127 g).*
Bottom: *Pyrrhotite, or iron sulfide, from the mines of Traversella, Piedmont, Italy.*

The mineral classification used in this book is a very simplified version of the mineralogical tables first established by Hugo Strunz in 1941 and most recently updated in 2001. It uses a chemical-structural classification in which the minerals are identified both on the basis of their chemical composition and their crystal structure. The system divides minerals into ten main classes; Strunz's tables further divide the minerals into a series of divisions, subdivisions and groups, but these have mostly been omitted here since they are not necessary to gain a basic understanding of the minerals.

Class 1: Native elements This class includes metals, semi-metals and non-metals.

Class 2: Sulfides and sulfosalts This class includes minerals that

present a formula of the type MxSz in which Mx contains metallic elements (iron, lead, copper) and Sz contains nonmetallic elements (sulfur and arsenic).

Class 3: Halides This class includes minerals containing halides, which are elements with particular chemical characteristics that include, for example, chlorine (Cl), fluorine (F) and bromine (Br).

Class 4: Oxides and hydroxides This class includes compounds in which

Right: *Fluorite is a typical halide mineral.*

oxygen combines with one or more metallic element; the subclass of the hydroxides includes those compounds in which hydrogen and oxygen are present in the form of OH or H_2O ions.

Class 5: Carbonates These present the combination of C (carbon) and O (oxygen) to form groups [CO_3]; other ions can also be present, such as OH and F.

Class 6: Borates These present the combination of B (boron) and O (oxygen) to form simple groups [BO_3] and far more complex arrangements.

Class 7: Sulfates, chromates, molybdates, tungstates They present S (sulfur), Cr (chromium), Mo (molybdenum) and W (tungsten), which combine with oxygen to form groups [SO_4], [CrO_4], [MoO_4] and [WO_4]; hydrogen and oxygen can also be present in the form of OH and H_2O ions.

Class 8: Phosphates, arsenates, vanadates These include P (phosphorus), As (arsenic) and V (vanadium), which combine with oxygen to form

Below: *This group of prismatic quartz crystals is a fine representative of the oxides.*

Below right: *Canavesite is a rare borate that was first found in the mines of Brosso, in Piedmont, Italy.*

Left: *This crystal of calcite is a fine example of the class of the carbonates.*

Below: *This bipyramidal crystal of scheelite is an example of the tungstates.*

groups [PO_4], [AsO_4] and [VO_4]; hydrogen and oxygen can also be present in the form of OH and H_2O ions, but other ions, such as fluorine and chlorine, are possible constituents.

Class 9: Silicates In this class, silicon (Si) combines with oxygen to form a tetrahedral group [SiO_4] and to create three-dimensional structures—independent tetrahedra, rings, chains or frameworks of tetrahedra—that are in turn divided into subclasses:

*Neosilicates: form independent tetrahedral [SiO_4] groups.

*Sorosilicates: [SiO_4] groups forming more or less complex frameworks represented by such formulas as [Si_2O_7].

*Cyclosilicates: [SiO_4] groups forming more or less complex rings represented by such formulas as [Si_6O_{18}].

*Inosilicates: [SiO_4] groups forming infinite chains, represented by such formulas as [Si_2O_6] and [Si_4O_{11}].

*Phyllosilicates: [SiO_4] groups forming infinite planar structures, represented by such formulas as [Si_2O_5].

*Tectosilicates: [SiO_4] groups forming complex three-dimensional frameworks.

Class 10: Organic minerals A series of minerals in which carbon combines with oxygen and hydrogen to form compounds of an organic nature. These are in turn divided into various subclasses, which involve the salts of organic acids, hydrocarbons and several organic mixtures of a highly complex nature.

Below: *Perfect prismatic crystal of diopside, a representative of the monoclinic pyroxenes, an inosilicate.*

NATIVE ELEMENTS

Class 1

METALLIC, SEMI-METALLIC AND NON-METALLIC MINERALS

Some minerals occur in the natural state in the form of native elements. These include the metallic elements, such as silver, gold, platinum and copper, which possess a high malleability and high thermal conductivity. Others, such as antimony, arsenic and bismuth—known as semi-metals—possess intermediate properties (for example, good thermal conductivity), but are somewhat fragile. The non-metallic minerals, including diamond, graphite and sulfur, present a variety of physical characteristics. Unlike the semi-metals, they do not have a metallic appearance and have a lower density.

Rare rhombohedral crystals of arsenic from Bohemia.

Antimony

1

3-3.5

6.61
6.72

Formula Sb
System Trigonal
Habit Appears in massive forms and granular aggregates; tin-white, with a metallic luster; its rare crystals are pseudocubic with somewhat rounded edges.
Environment Found in hydrothermal veins in association with other minerals of antimony.
Name and notes From the Latin *antimonium* ("monk's bane"), used by the medieval scholar Constantine the African circa A.D. 1050 to describe the mineral stibnite, since it was the cause of frequent, and in some cases fatal, poisonings among monks who ate with utensils made of this metal.

Specimen of antimony from Sarawak, Borneo.

Arsenic

2

Formula As
System Trigonal
Habit Usually forms massive or concentric mammillary aggregates and stalactites; light gray, with a metallic luster, although the mineral tarnishes quickly to dark gray; rare crystals are rhombohedral and millimetric in size.
Environment Found in hydrothermal veins associated with other minerals of arsenic and in deposits containing minerals of silver and cobalt.
Name and notes Already known in 400 B.C.; the name is from the Greek *arsenikos* or *arsenikon*, meaning "male" and thus strong, a reference to its potent properties, including that of being a strong poison.

Specimen of arsenic from Germany showing the characteristic mammillary form.

Bismuth

3

2-2.5
9.70
9.83

Formula Bi
System Trigonal
Habit Forms granular, reticulate or arborescent masses; silver-white, with a metallic luster; there have been instances of prismatic crystals, isolated or in groups, several inches long. Such crystals are usually inclusions in rock, although there are exceptional distinct crystals.
Environment Found in hydrothermal veins, where it is associated with minerals of silver, cobalt, nickel and tin; may also occur in granitic pegmatites and in veins of quartz containing topaz, minerals of tin and members of the wolframite series.
Name and notes The name refers to the word *wisimut*, from the Old German *wis mat*, "white mass," an allusion to the silvery white reflections of the mineral.

Specimen of bismuth from Saxony.

Copper

4

2.5-3 8.95

Formula Cu
System Cubic
Habit Forms odd crystals of rhombododecahedral, cubic and even more complex habits; copper-red with a metallic luster; dendritic and filiform aggregates are common, as are masses weighing up to several tonnes.
Environment Commonly associated with basic igneous rocks, it is also found in superficial zones of metallic deposits, where it forms through the process of oxidation of minerals containing copper.
Name and notes This metal has been used since the Bronze Age; the name is from the Latin *cuprum* or *cyprium*, as it was believed to come from the island of Cyprus.

Specimen of distinct copper crystals from Michigan.

Diamond

5

10 3.51

Diamond crystal from South Africa.

Formula C
System Cubic
Habit Forms interesting octahedral, rhombododecahedral, cubic and even more complex crystals; perfectly colorless or yellow, brown or black, and rarely pink, green, blue, orange or red, with an adamantine luster.
Environment Diamond crystals are found in so-called kimberlite pipes, volcanic conduits composed of intrusive ultrabasic rock, originating at great depths in the earth's upper mantle. Given their extreme hardness and resistance to erosion, diamond crystals are also found in alluvial deposits, results of the erosion of such pipes.
Name and notes Probably derived from the Greek *adamas*, "invincible," an allusion to its extreme hardness.

Gold

6

2.5-3 | 19.30

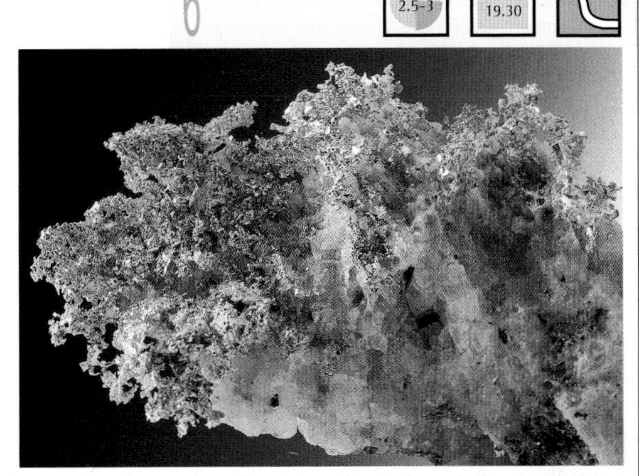

Formula Au
System Cubic
Habit Forms octahedral, dodecahedral, cubic or even more complex crystals of characteristic gold-yellow color, with a metallic luster; also occurs in dendritic, filiform or spongelike aggregates, or in scales, lumps and nuggets, in some cases weighing several pounds.
Environment Found in deposits in quartz veins of hydrothermal origin, associated with pyrite and other sulfides; also present in some contact-metamorphic rocks and common, in variable concentrations, in alluvial deposits.
Name and notes The etymology of the name has been lost, but the name is believed to derive from the Sanskrit *jyal* and the German *geld*.

Typical example of gold from Brusson, Valle d'Aosta, Italy.

Graphite

7

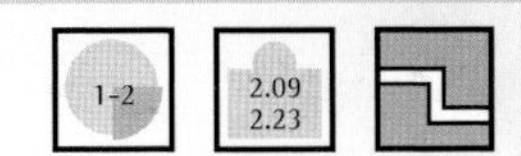

Formula C
System Hexagonal
Habit Most common in massive, columnar, granular or earthy masses; black, with dull metallic luster; crystals have a well-defined hexagonal habit and spheroidal aggregates are rare.
Environment Results from the metamorphism of carbon-rich sedimentary rocks; is also found as the primary mineral in some igneous rocks.
Name and notes From the Greek *graphein*, "to write," an allusion to its ancient use as a drawing pencil. Graphite is the low-pressure polymorph of carbon, having the same chemical composition as diamond, but a different crystalline structure.

Specimen of graphite from Sri Lanka.

Mercury

8

n.d. 13.60

Formula Hg
System Trigonal at –38.9°C
Habit Occurs as liquid spheres; tin-white, with a metallic luster.
Environment Found in low-temperature hydrothermal deposits related to active geothermal activity.
Name and notes Known since at least 1500 B.C.; its name is from the Latin *mercurius*, the term used by alchemists for this metal, and also from the Latin *hydrargyrum*, "quicksilver."

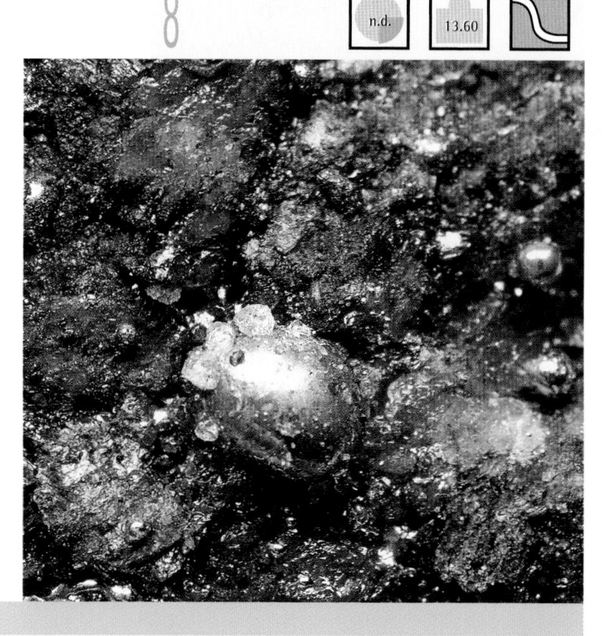

Mercury from Levigliani, Tuscany, Italy.

Platinum

9

4-4.5 14.00 19.00

Formula Pt
System Cubic
Habit Rare metal in nature; even more rare are its cubic crystals; generally found in granular form or in nuggets, which on occasion can reach 18 to 20 pounds (8 to 9 kg) in weight; steel-black, with metallic luster.
Environment Almost always found in alluvial deposits, although it originates in intrusive ultrabasic igneous rocks.
Name and notes From the Spanish *platina del Pinto*, "silver of the Pinto," since it was discovered near the Pinto River in Colombia.

Platinum nugget weighing 1½ ounces (50 g) from alluvial deposits in Russia.

Silver

10

2.5–3 | 10.10 11.10

Formula Ag
System Cubic
Habit Usually found in groups of cubic octahedral or dodecahedral crystals, or in filiform or reticular arborescent forms; characteristically silvery white in color, with a metallic luster, tarnishes to gray or black.
Environment Found in hydrothermal veins containing quartz; also originates through secondary processes of oxidation in deposits containing minerals rich in silver.
Name and notes From the Anglo-Saxon *siolfor,* the exact meaning of which has been lost.

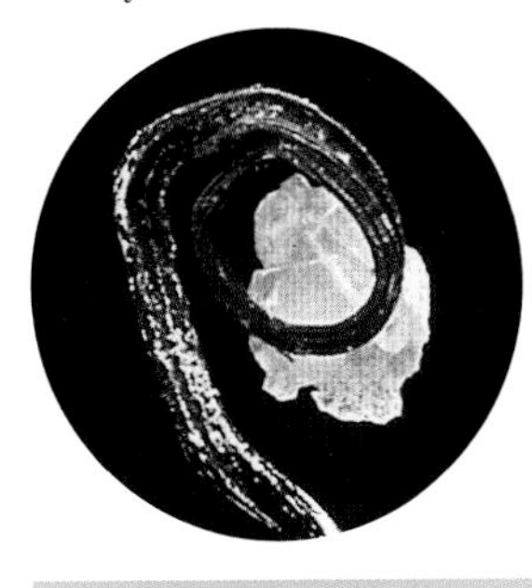

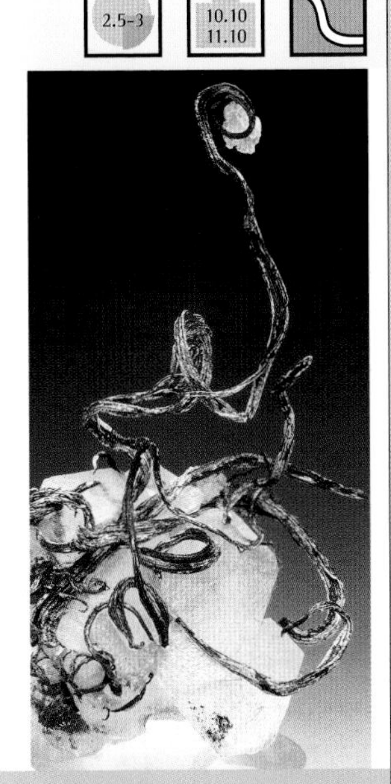

Curls of filiform silver on calcite from Sarrabus, Sardinia, Italy.

Sulfur

11

1.5–2.5 | 2.07

Formula S
System Orthorhombic
Habit Forms beautiful crystals with bipyramidal habit; yellow, brownish yellow, or reddish yellow, with resinous luster; also common in massive, mammillary and stalactitic forms.
Environment Characteristic product of volcanic sublimation, and found in sedimentary rocks subject to the activity of micro-organisms; forms from the decomposition of sulfides, through chemical reactions caused by the circulation of highly acidic water in deposits rich in sulfides.
Name and notes Known since at least 2000 B.C.; it refers to the chemical element and is based on the Latin *sulphurum.*

Typical bipyramidal crystals of sulfur from the Marches, Italy.

SULFIDES AND SULFOSALTS

Class 2

PYRITE GROUP

Nearly 20 minerals belong to the pyrite group. These sulfides possess a cubic symmetry and a general formula of the type AX_2, in which A can contain cobalt, iron, manganese, nickel, copper, gold, palladium or platinum, whereas X_2 can be antimony, arsenic, bismuth, selenium or tellurium.

Pyrite, for which the pyrite group is named, can appear in marvelous, perfectly formed crystals, as in this specimen from a mine at Niccioleta, Tuscany, Italy.

Hauerite

12

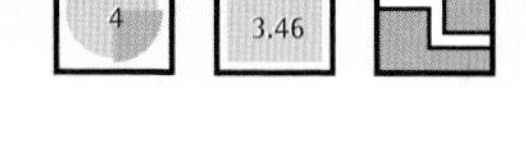

Formula MnS_2
System Cubic
Habit Forms octahedral or cubo-octahedral crystals; brownish black, with a metallic luster that is usually not visible because of tarnishing.
Environment Occurs at low temperatures and is associated with the circulation of sulfurous fluids within clay-rich sedimentary rocks.
Name and notes Named after the Austrian geologists Joseph von Hauer (1778–1863) and his son Franz von Hauer (1822–99); discovered near Kalinka in the Czech Republic.

Cubo-octahedral crystal of hauerite from Raddusa, Sicily.

Pyrite

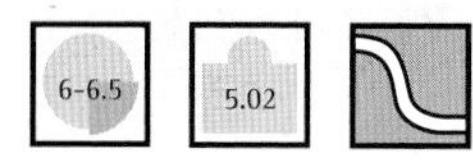

Formula FeS_2
System Cubic
Habit Forms cubic, octahedral and pyritohedral crystals, as well as combinations of those forms; pale to brass-yellow, with a metallic luster; crystals commonly have parallel striations and are twinned by penetration; also common as granular masses, and acicular, mammillary and stalactitic aggregates are not rare.
Environment Characteristic of hydrothermal veins, also occurs as an accessory of igneous and pegmatitic rocks; present in many metamorphic and sedimentary rocks.

Name and notes From the Greek *pyr*, "fire," an allusion to its property of giving off sparks if struck with a piece of steel.

Cubic crystals of pyrite with calcite and sulfur from the mine at Niccioleta, Tuscany, Italy.

Sperrylite

14

6-7 10.58

Formula $PtAs_2$
System Cubic
Habit Forms cubic or cubo-octahedral crystals; tin-white with a black streak and distinct metallic luster.
Environment Occurs in platiniferous deposits associated with intrusive ultrabasic igneous rocks.
Name and notes Named after the American chemist Francis Lewis Sperry (1861–1906), who discovered it. First found at the Vermilion mine, near Sudbury, Ontario, Canada.

Sperrylite crystal from Talnakh, Russia.

Acanthite (Argentite)

15

2-2.5 | 7.22

Formula Ag_2S
System Monoclinic
Habit Usually found in groups of pseudocubic, pseudo-octahedral or prismatic crystals; lead-gray to black, with a metallic luster.
Environment Forms in hydrothermal veins with quartz, associated with other sulfides containing silver, and through alteration of primary sulfides containing silver.
Name and notes From the Greek *akantha,* "spiny," the shape of its crystals, and the Latin *argentum,* "silver." The higher-temperature polymorph argentite transforms to acanthite at –279°F (–173° C), maintaining the same chemical composition but assuming a different structure.

Acanthite specimen from Mexico.

Arsenopyrite

16

5.5-6 | 6.07

Formula FeAsS
System Monoclinic
Habit Forms prismatic or tabular crystals, commonly striated and twinned; silver-white or steel-gray, with metallic luster; also common in massive and granular forms.
Environment Forms in hydrothermal veins rich in quartz, where it is associated with numerous other sulfides; also found in pegmatites and metamorphic rocks, in particular gneisses, mica schists and skarns.
Name and notes Known in antiquity as "arsenical pyrites" for its resemblance to pyrite and its content of arsenic, of which *arsenopyrite* is a contraction.

Crystals of arsenopyrite associated with quartz from Mexico.

Bismuthinite

17

2-2.5 | 6.78

Formula Bi_2S_3
System Orthorhombic
Habit Commonly occurs in massive form but also in prismatic crystals, acicular, elongate and striated habits; lead-gray to tin-white, with a metallic luster.

Environment Found in hydrothermal veins; also in granitic pegmatites, in deposits associated with metamorphic rocks, and as the result of sublimation at fumaroles.

Specimen with prismatic crystals from Bolivia.

Name and notes The name reflects its elevated content of bismuth.

Bornite

18

3-3.25 | 5.06 5.08

Formula Cu_5FeS_4
System Orthorhombic
Habit Found in compact, granular masses; copper-red or violet, with superficial blue iridescence and metallic luster; less common as pseudocubic and pseudododecahedral crystals; commonly twinned by penetration.
Environment Occurs disseminated in certain mafic igneous rocks; common in hydrothermal and skarn-type deposits; also present in clay-rich sedimentary rocks containing copper minerals.
Name and notes Named after the Austrian mineralogist, paleontologist and metallurgist Ignaz von Born (1742–91).

Bornite crystals with calcite from Dalnegorsk, Russia.

Calaverite

19

Formula $AuTe_2$
System Monoclinic
Habit Rare prismatic crystals, striated along the faces, usually millimetric in size; characteristically brass-yellow, with a strong metallic luster; also present in massive and granular forms.
Environment Found in hydrothermal environments in quartz veins and associated with other rare minerals of tellurium, gold and mercury.
Name and notes Named after Calaveras, California, where it was discovered at the Stanislaus mine.

Typical calaverite crystals from Colorado.

Carrollite

20

Formula $Cu(Co,Ni)_2S_4$
System Cubic
Habit Forms beautiful cubo-octahedral crystals; steel-gray, with a strong metallic luster; also common in massive and granular forms.
Environment Characteristic of hydrothermal deposits, where it is associated with numerous other sulfides containing cobalt, nickel, iron and copper.
Name and notes Named after Carroll County, Maryland, where it was found in the Patapsco mine in Finksburg.
Crystal from the Democratic Republic of Congo.

Chalcocite

21

2.5-3

5.50
5.80

Formula Cu_2S
System Monoclinic
Habit Found in massive forms, granular and compact, but not infrequent as prismatic crystals, tabular or twinned; lead-gray, with a metallic luster.
Environment Secondary mineral typical of the oxidation zones of porphyry copper deposits; more rarely forms as primary mineral in hydrothermal veins.
Name and notes From the Greek *chalkos*, "copper," with reference to the high contents of that element in the mineral.

Chalcocite crystal from the Number 57 mine, Kazakhstan.

Chalcopyrite

22

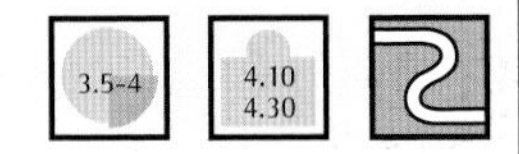

Tetrahedral crystals of chalcopyrite from a mine at Trepca, Kosovo.

Formula $CuFeS_2$
System Tetragonal
Habit Common in massive and compact forms; very characteristic in tetrahedral and scalenohedral crystals, commonly twinned by penetration; brass-yellow, generally iridescent, with a metallic luster.
Environment Somewhat common in hydrothermal veins, associated with other sulfides of copper; also occurs as an accessory mineral in mafic intrusive igneous rocks.
Name and notes From the Greek words *chalkos*, "copper," referring to its copper content, and *pyrites*, "fiery."

Cinnabar

23

2-2.5 | 8.18

Formula HgS
System Trigonal
Habit Very attractive rhombohedral crystals; bright red, in some cases with brownish hues, with an adamantine luster; also forms large microcrystalline incrustations and granular masses.
Environment Occurs in low-temperature hydrothermal solutions, where it is associated with mercury; also found as an impregnation in igneous, metamorphic and sedimentary rocks.
Name and notes Probably from an ancient Arab word, *zinjafr*, or the Persian *zinjifrah*, "dragon's blood," in allusion to its color.

Splendid crystals from the Tao mine, Hunan region, People's Republic of China.

Covellite

24

1.5-2 | 4.60 4.76

Formula CuS
System Hexagonal
Habit Common in massive and laminar forms; crystals, invariably rare, are hexagonal and tabular; highly iridescent in blue-red or brass-yellow, with a metallic luster.
Environment Secondary mineral found in zones of oxidation of deposits rich in sulfides of copper; occurs more rarely as a primary sulfide in hydrothermal veins.
Name and notes Named after the Italian mineralogist Niccolò Covelli (1790–1829), who discovered it on Mount Somma-Vesuvius, near Naples, Italy.

Iridescent crystals with pyrite from Calabona, Sardinia, Italy.

Enargite

25

3 | 4.45

Formula Cu_3AsS_4
System Orthorhombic
Habit Typically forms tabular or prismatic crystals, strongly striated; gray, gray-black, commonly iridescent, with a metallic luster.
Environment Occurs in hydrothermal veins associated with numerous other sulfides of lead, copper and zinc.
Name and notes From the Greek *enarge*, "visible," an allusion to its perfect cleavage.

Enargite crystals from Furtei, Sardinia, Italy.

Galena

26

2.50 2.75 | 7.58

Formula PbS
System Cubic
Habit Common in cubic crystals, also twinned, cubo-octahedral, and more rarely octahedral; lead-gray, with a metallic luster; also found in massive forms with clear cleavage planes and in reticulate, skeletal and granular masses.
Environment Very common sulfide, found in hydrothermal veins, and also present in metamorphic rocks, in granitic pegmatites and in sedimentary rocks, particularly limestone and dolomite.
Name and notes From the Latin *galena*, used for the dross obtained from the process of melting lead.

Cubic crystals of galena from Madan, Bulgaria.

Greenockite

27

3-3.5 4.82

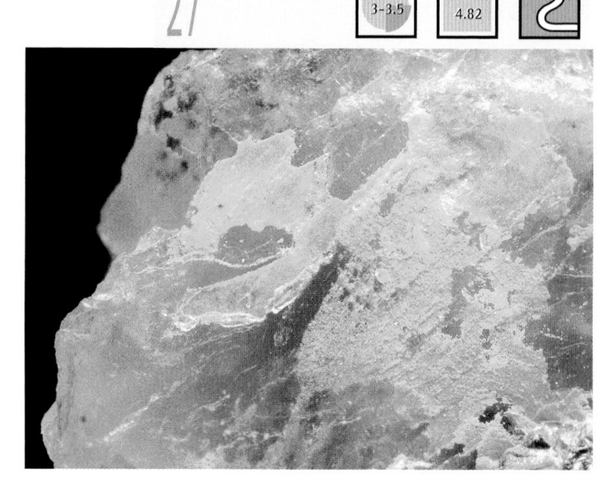

Formula CdS
System Hexagonal
Habit Most common as an earthy powder or microcrystalline incrustation; yellow or orange, with an adamantine to resinous luster; very rare in crystals with a pyramidal form.
Environment Occurs as a product of the alteration of sphalerite; the crystals, always very rare, have been found in cavities in intrusive igneous rocks.
Name and notes Named after the British official and dedicated collector Lord Greenock (Charles Murray Cathcart, 1783–1859), who discovered it. First found in Bishopton, Scotland.

Cadmium yellow powdery greenockite on fluorite, Dossena mine, Lombardy, Italy.

Kermesite

28

Formula Sb_2S_2O
System Triclinic
Habit Forms very elongate prismatic crystals or radiating groups; dark red, with an adamantine luster.
Environment Secondary mineral, found in hydrothermal formations from the alteration of stibnite; commonly associated with other minerals of antimony.
Name and notes An ancient name from the Greek *kermes* or the Persian *qurmiza*, both used by ancient alchemists to indicate the powder of the oxide of antimony, with its typical red color.

Centimetric prismatic crystals from Guangxi Province, People's Republic of China.

Marcasite

29

Formula FeS_2
System Orthorhombic
Habit Common in tabular crystals, bipyramidal to prismatic, with typical curving faces; bronze-yellow, bronze or tin-white, with a metallic luster; also occurs in massive, granular and stalactitic forms.
Environment Deposited at low temperatures in hydrothermal veins; accessory in igneous rocks and pegmatites; widespread in many metamorphic rocks and somewhat common in sedimentary rocks.
Name and notes From the Assyrian *Markhashitu,* synonym of the Arabic *Markhashi,* the name of an ancient Persian province that was an early source of the mineral. Marcasite is dimorphous with pyrite.

Minute crystals of marcasite cover a crystal of calcite from the United States.

Millerite

30

3-3.5
5.50

Formula NiS
System Trigonal
Habit Forms isolated crystals and radiating groups or tufts with a typical, very slender capillary appearance, thin and flexible; brass-yellow or bronze-yellow, with a metallic luster.
Environment Found in cavities of carbonate-rich sedimentary rocks, deposited by the circulation of low-temperature hydrothermal fluids; also constitutes a typical product of alteration in minerals of nickel in serpentinites.
Name and notes Named after the British mineralogist William Hallowes Miller (1801–80), author of an important book on crystallography.

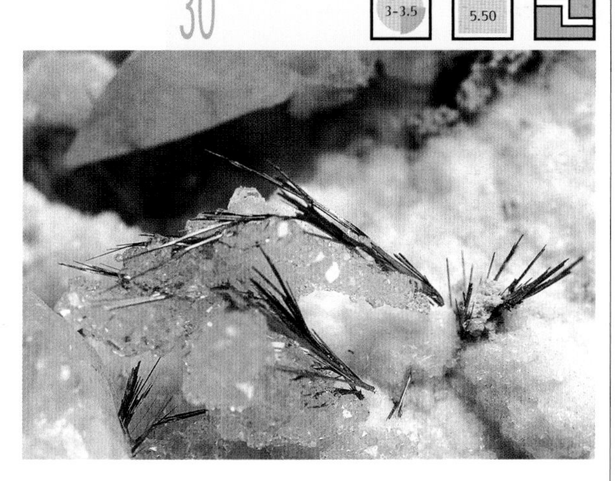

Millerite on quartz from the Apennines near Bologna, Italy.

Molybdenite

31

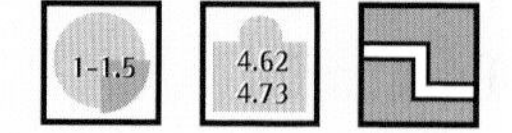

Formula MoS_2
System Hexagonal
Habit Forms tabular crystals with hexagonal contours, laminar and flexible; lead-gray, with a strong metallic luster; also found in foliated masses and as scales.
Environment Occurs in hydrothermal quartz veins and in porphyry copper deposits; also found in metamorphic rocks and in intrusive rocks, in particular granites, aplites and pegmatites.
Name and notes From the Greek *molybdos*, "lead," a result of ancient confusion among galena, graphite and molybdenite.

Specimen of molybdenite from Ontario, Canada.

Orpiment

32

1.5-2

3.49

Formula As_2S_3
System Monoclinic
Habit Common in foliated, fibrous or columnar aggregates; more rarely in prismatic crystals; lemon-yellow or yellowish brown, with resinous or pearly luster.
Environment Occurs in low-temperature hydrothermal veins; also occurs as a typical product of sublimation in fumaroles and formed through the alteration of minerals of arsenic, especially realgar.
Name and notes From the Latin words *aurum*, "gold," and *pigmentum*, "color," an allusion to its color and also to the ancient belief that this mineral could contain gold.

Crystals of orpiment from Nevada.

Pyrrhotite

33

Formula $Fe_{1-x}S$
System Monoclinic or hexagonal
Habit Tabular crystals, flattened to form aggregates of parallel pseudopyramidal crystals and groups of crystals joined in rosettes; bronze-yellow, reddish bronze, with a metallic luster; common in massive and granular forms.
Environment Found in mafic igneous rocks and in pegmatites; also present in metamorphic skarn and in carbonate-rich sedimentary rocks.
Name and notes From the Greek *pyrrotes*, "reddish," referring to the pale reddish reflections characteristic of this mineral.

Rosette of tabular hexagonal crystals from Dalnegorsk, Russia.

Realgar

34

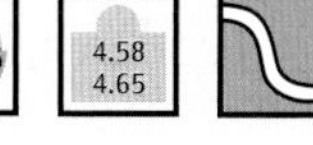

Formula As_4S_4
System Monoclinic
Habit Forms striated prismatic crystals; common in aggregates of massive appearance, granular and microcrystalline; red or orange, with a resinous luster.
Environment Occurs in hydrothermal veins, where it is associated with other minerals of arsenic and antimony; a typical product of volcanic sublimation, it is also found in carbonate- and clay-rich sedimentary rocks.
Name and notes From the ancient Arabic *rahj al-gar*, "powder of the mine," because it was frequently found associated with minerals of silver.

Striated prismatic crystals of realgar from the People's Republic of China.

Skutterudite

35

5.5-6 | 6.50

Formula $CoAs_3$
System Cubic
Habit Forms cubic, octahedral and rhombododecahedral crystals; tin-white to silver-gray, with a metallic luster; common in massive and granular forms; also found in skeletal aggregates.
Environment Characteristic mineral of hydrothermal veins, where it is associated with minerals of nickel and cobalt.
Name and notes Named after Skutterud, Norway, where it was discovered.

Cubo-octahedral crystal of skutterudite from Morocco.

Sphalerite

Formula ZnS
System Cubic
Habit Forms tetrahedral, dodecahedral and more complex crystals with typical curved faces; colored from red to green to yellow or (more commonly) brown, gray or black, with resinous or adamantine luster.
Environment Typical of hydrothermal formations, where it is associated with numerous other minerals, in particular sulfides and carbonates; also present in deposits of sedimentary carbonate-rich rocks, in which it can form large deposits.
Name and notes From the Greek *sphaleros,* "treacherous," because in the past it was often mistaken for other minerals, notably galena.

Crystal of sphalerite from Carrara, Tuscany, Italy.

Stibnite

37

2 | 4.63

Formula Sb_2S_3
System Orthorhombic
Habit Forms typical elongate, prismatic crystals; lead-gray, with a metallic luster; bent crystals and radiating aggregates formed of very thin elongated crystals are common.
Environment Forms in hydrothermal veins, usually associated with other minerals containing antimony.
Name and notes From the Greek name for the mineral, *stibi,* from which came the Latin term *stibium,* or "cane."

Striated prismatic crystals of stibnite from Romania.

Wurtzite

38

3.5–4 | 4.09

Formula ZnS
System Hexagonal
Habit A somewhat rare form of ZnS, which forms pyramidal crystals, striated horizontally; brownish red, brownish yellow, brown or brownish black, with a resinous or adamantine luster.
Environment Formed in hydrothermal veins, and found in metamorphic rocks, in particular marble and skarn.
Name and notes Named after the French chemist Charles Adolphe Wurtz (1817–84), dean of the medical school at the University of Paris. Wurtzite is dimorphous with sphalerite, meaning it has the same chemical composition but has a different crystalline structure.

Typical hexagonal pyramidal crystal of wurtzite from marble in Carrara, Tuscany, Italy.

TETRAHEDRITE GROUP

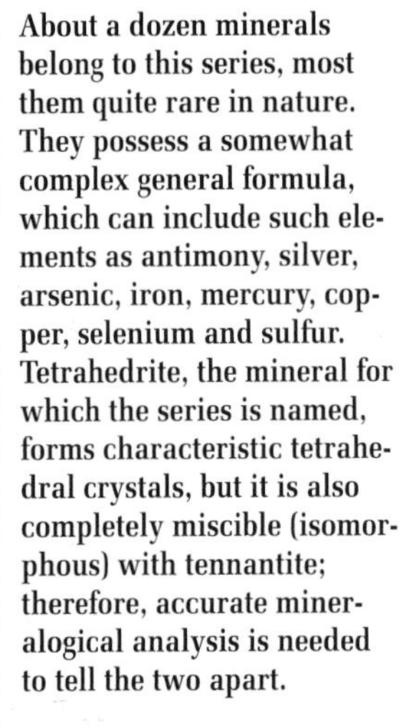

About a dozen minerals belong to this series, most them quite rare in nature. They possess a somewhat complex general formula, which can include such elements as antimony, silver, arsenic, iron, mercury, copper, selenium and sulfur. Tetrahedrite, the mineral for which the series is named, forms characteristic tetrahedral crystals, but it is also completely miscible (isomorphous) with tennantite; therefore, accurate mineralogical analysis is needed to tell the two apart.

Crystals of tetrahedrite from Schwaz in Austria.

Tennantite

39

3-4.5 | 4.62

Formula $Cu_6Cu_4(Fe_1Zn)_2(As_1Sb)_4S_{13}$
System Cubic

Habit Commonly forms tetrahedral crystals, in some cases twinned by penetration, commonly joined in groups of parallel crystals; gray or black, with a metallic luster; also common in massive and granular forms.

Environment Occurs together with other sulfides of arsenic, iron and copper in quartz-bearing hydrothermal veins and in metamorphic skarn.

Name and notes Named after the English chemist Smithson Tennant (1761–1815), who discovered the chemical elements iridium and osmium.

Tennantite crystals from Casapalca, Peru.

Tetrahedrite

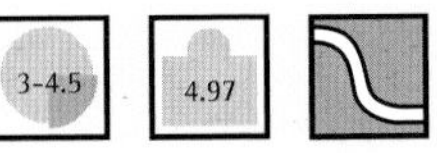

Formula $Cu_6Cu_4(Fe_1Zn)_2(Sb_1As)_4S_{13}$
System Cubic

Below: Tetrahedrite crystals with fluorite from Saxony.

Bottom: Tetrahedrite crystal with complex habit from the Binn Valley, Switzerland.

Habit Like tennantite, forms tetrahedral crystals, in some cases twinned by penetration, commonly joined in groups of parallel crystals; gray or black, with metallic luster; also common in massive or granular forms.
Environment Found together with other sulfides in hydrothermal veins of quartz and in metamorphic skarn.
Name and notes Named for the form of its crystals. It has the same crystalline structure as tennantite but the composition is enriched in antimony rather than arsenic.

Boulangerite

41

2.5–3 6.20

Formula $Pb_5Sb_4S_{11}$
System Monoclinic
Habit Forms very thin acicular crystals and compact fibrous masses; lead-gray, with metallic luster.
Environment Found in hydrothermal quartz veins and in metamorphic skarn, associated with other sulfides of iron, lead and copper.
Name and notes Named after the French mining engineer Charles Boulanger (1810–49), who was the first to describe it.

Acicular crystals of boulangerite from the Brosso mine, Piedmont, Italy.

Bournonite

42

Formula $PbCuSbS_3$
System Orthorhombic
Habit Forms prismatic to tabular crystals, usually striated; also common in aggregates of twinned crystals forming a wheel; steel-gray, with a metallic luster.
Environment Found in hydrothermal quartz veins and in metamorphic skarn, associated with other sulfides of copper and lead.
Name and notes Named after French scholar and mineral collector Count Jacques Louis de Bourbon (1751–1825). It was first found in the Wheal Boys mine in Cornwall, England.

Twinned crystal of bournonite associated with dolomite, Brosso mine, Piedmont, Italy.

Proustite

43

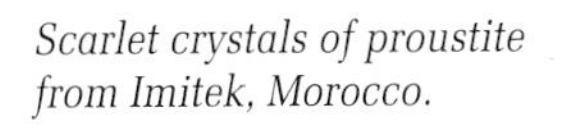

Formula Ag_3AsS_3
System Trigonal
Habit Forms prismatic rhombohedral or scalenohedral crystals; scarlet-vermilion but darkens on exposure to light; strong adamantine luster; also present in massive and granular forms.
Environment Found in hydrothermal quartz veins and in the oxidation zones of ore deposits, in association with other minerals containing silver, arsenic and antimony.
Name and notes Named after the French chemist Joseph Proust (1755–1826).

Scarlet crystals of proustite from Imitek, Morocco.

Pyrargyrite

44

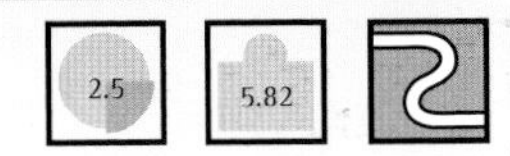

Formula Ag_3SbS_3
System Trigonal
Habit Common in massive and granular forms; also present in prismatic crystals, pseudohexagonal as a result of cyclic twinning; bright red or black, with an adamantine luster.
Environment Forms in hydrothermal veins, both as a primary mineral and the product of the alteration of minerals containing silver and antimony.
Name and notes From the Greek *pyr*, "fire," and *argyros*, "silver," a reference to its red color and silver content.

Typical crystals of pyrargyrite from Zacatecas, Mexico.

Class 3

Atacamite

45

Formula $Cu_2(OH)_3Cl$
System Orthorhombic
Habit Forms slender prismatic crystals, striated and joined in radiating groups; emerald-green, with a vitreous luster; also occurs in massive aggregates, both fibrous and granular.
Environment Formed through the oxidation of other minerals of copper, especially in arid environments having high concentrations of salts; also found in fumaroles. This mineral appears as a product of alteration in ancient archeological objects made of bronze and copper.
Name and notes Named after Atacama, Chile, near where it was discovered.

Prismatic crystals of atacamite from Atacama, Chile.

Boleite

46

3-3.5

5.05

Formula
$KPb_{26}Cu_{24}Ag_9(OH)_{48}Cl_{62}$
System Cubic
Habit This fairly rare mineral forms beautiful cubic crystals; indigo-blue, greenish blue, with a vitreous luster.
Environment Secondary mineral that originates from the reaction among sulfides of silver, lead and copper from saline fluids; also forms in vesicular cavities in scoria in contact with saltwater.
Name and notes Named after Boleo in Baja California, where it was discovered.

Cubic crystal of boleite from Baja California, Mexico.

Chlorargyrite

47

2-2.5 5.55

Formula AgCl
System Cubic
Habit Rare mineral that forms cubic crystals, usually joined in parallel aggregates; pearl-gray, brown or colorless, with resinous to adamantine luster; more commonly forms thin microcrystalline incrustations and coralloid or stalactitic aggregates.
Environment Forms in arid areas with high temperatures through the alteration and oxidation of minerals containing silver.
Name and notes Named for its chemical composition of chlorine and silver (Greek *argyros*, "silver"). It changes color when exposed to sunlight, becoming brown or violet.

Chlorargyrite crystal from Broken Hill, Australia.

Cryolite

48

2.5 2.97

Formula Na_3AlF_6
System Monoclinic
Habit Common in massive or granular forms, more rarely as pseudocubic crystals; colorless, white or reddish, with a vitreous luster.
Environment Mineral characteristic of alkaline granites and related pegmatites; also occurs in rhyolitic rocks in close association with fluorite and topaz; also present in carbonatites, products of crystallization of a carbonate magma.
Name and notes From the Greek words *kryos*, "ice" and *lithos*, "stone," an allusion to its appearance. The mineral was first found near Ivigtut, Greenland, where it is abundant and was extracted for industrial use.

Cubic crystals of cryolite from Ivigtut, Greenland.

Fluorite

49

4 | 3.18

Formula CaF_2
System Cubic
Habit Forms beautiful cubic, octahedral and rarely dodecahedral crystals, or combinations of complex forms, in a wide variety of colors (violet, blue, green, yellow, pink, white), with a vitreous luster; also common in massive and granular forms, commonly with variations in color.
Environment Very common mineral in hydrothermal veins; also present in metamorphic rocks, in pegmatites and in sedimentary rocks—most of all in dolomitic limestone, where it can form large deposits.
Name and notes From the Latin *fluere*, "to flow," reflecting the fact that it melts more easily than other minerals with which it was confused in the past.

Octahedral fluorite crystal from Italy's Ossola Valley.

Halite

50

2-2.5 | 2.17

Formula NaCl
System Cubic
Habit Forms crystals that are generally cubic; colorless, gray, yellow, orange, pink, blue or violet, with a vitreous luster; also common in skeletal and granular stalactitic aggregates.
Environment Forms in evaporite deposits of sedimentary rocks, sometimes forming beds of enormous thickness; also occurs as a product of sublimation in fumaroles.
Name and notes From the Greek *hals,* "salt"; halite is commonly known as rock salt or table salt.

Halite crystal from Heringen, Germany.

Sylvite

51

2

1.99

Formula KCl
System Cubic
Habit Forms compact granular masses and also stalactitic forms; less common as cubic crystals or combinations of cube and octahedron; colorless, white, bluish or reddish yellow through the inclusion of hematite, with a vitreous luster.
Environment Forms in evaporite sequences in sedimentary basins associated with halite; also occurs as a product of sublimation in fumaroles.
Name and notes From the Latin *sal digestibus Sylvii,* "digestive salt of Sylvius"; its therapeutic uses were first recognized by the Dutch physicist and chemist François Sylvius de la Boe (1614–72). The mineral was first discovered in the fumaroles of Mount Somma-Vesuvius, Italy.

Small cubic crystals of sylvite from Wieliczka, Poland

Villiaumite

52

Formula NaF
System Cubic
Habit Forms massive, granular or flaky crystals, rarely forms cubic crystals; carmine-red or orange, with a vitreous luster.
Environment Found in intrusive igneous rocks, in particular nepheline syenites and related alkaline pegmatites.
Name and notes Named after Maxime Villiaume, an officer in the French colonial army stationed in Madagascar. Villiaume helped the famous mineralogist Alfred Lacroix augment the collections of minerals from Madagascar in French museums. The mineral was first discovered on Rouma Island, Equatorial Guinea. It tends to quickly alter upon exposure to air.

Crystal with cubic cleavage from Equatorial Guinea.

OXIDES
AND HYDROXIDES

HEMATITE GROUP

This group includes four minerals; the two illustrated here are of particular interest because of their industrial uses. These oxides belong to the trigonal system and have the general formula of the type R_2O_3 (R = Al, Cr, Fe, V) and have somewhat high densities. Corundum and hematite can form beautiful crystals that are highly prized by collectors.

Aggregate of hematite crystals from Brazil.

Corundum

53

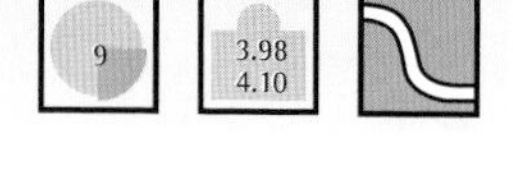

Formula Al_2O_3
System Trigonal
Habit Forms prismatic crystals with pseudohexagonal contours, generally striated, and less common are bipyramidal and rhombohedral crystals; occurs in the entire gamut of colors, with a luster from vitreous to adamantine.

Environment Forms in metamorphic rocks, like marbles and eclogites, rich in aluminum and in intrusive igneous rocks, such as monzonites, syenite and alkaline pegmatites.
Name and notes Known since antiquity, the name stems from *kuruntam*, a Tamil word based on the Sanskrit *kuruvinda*, probably meaning "ruby." Corundum is used to make abrasives and refractories; transparent colored varieties are cut to produce gemstones of great value, including red rubies and sapphires.
Corundum from Afghanistan.

Hematite

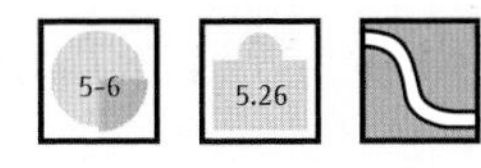

Formula Fe_2O_3
System Trigonal
Habit Forms many-faceted rhombohedral crystals with typical triangular striations, or laminar crystals joined in rosettes; also in botryoidal and stalactitic forms; steel-gray, with a metallic luster.
Environment Widely distributed in intrusive and metamorphic rocks, including banded iron formations; typical product of sublimation along volcanic conduits; also present in sedimentary rocks.
Name and notes Referring to this mineral in 315 B.C., Theophrastus used the Greek word *haimatites*, "blood-colored stone," because of the vivid red color of its powder. The mineral has many industrial uses, including in the manufacture of steel.

Below: Specimen of hematite from Mount Vesuvius, Italy. Bottom: Classic crystal of hematite from the island of Elba, Italy.

PEROVSKITE GROUP

Very few species belong to this group, and those illustrated here, although among the most common, are not easily found. These oxides form cubic and pseudocubic crystals and have a general formula of the type ABO_3 (A = Ca, Ce, Na, Sr; B = Nb, Ti, Fe), with somewhat high density.

The minerals of this group are of great scientific interest, as they form high concentrations valued as a source of rare elements.

Pseudocubic crystals of perovskite from the Susa Valley, Italy.

Loparite-(Ce)

55

Formula $(Ce,Na,Ca)(Ti,Nb)_3$
System Orthorhombic, pseudocubic
Habit Forms characteristic cubic and octahedral crystals up to ¾ inch (2 cm) across; black or dark gray, with a metallic luster.
Environment Found exclusively in certain intrusive igneous rocks, including nepheline syenites, alkaline pegmatites and carbonatites.
Name and notes From the Russian *Lopar,* "Laplander," since it was first found in the tundra of Russian Lapland. A natural source of several rare elements, including niobium, tantalum, titanium and the rare earths, loparite-(Ce) is of great scientific interest.

Typical twinned crystal of loparite-(Ce) from the Kola Peninsula, Russia.

Perovskite

5.5

3.98
4.26

Formula $CaTiO_3$
System Orthorhombic, pseudocubic
Habit Forms characteristic crystals that resemble distorted cubes, less commonly cubo-octahedral, and more rarely in skeletal, dendritic aggregates; black, brown, reddish brown or yellow, with an adamantine to metallic luster.
Environment Found as an accessory in certain alkaline igneous rocks, such as nepheline syenites, kimberlites and carbonatites; also present in such metamorphic rocks as serpentinites and rodingites.
Name and notes Named after Count Lev Aleksevich von Perovsky (1792–1856), Russian official and mineral agent. The mineral was first found near Slatoust in the Ural Mountains of Russia.

Top: Perovskite crystal from Emarese, Valle d'Aosta, Italy. Above: Pseudocubic crystals from the Susa Valley, Italy.

PYROCHLORE GROUP

More than 20 minerals belong to this group; with the exception of the three illustrated here, they are all very rare. These complex oxides present a cubic symmetry and the crystals, generally tiny in size, typically have an octahedral habit. Their chemical formulas are very complex, and where they are present in high concentrations they are of great strategic interest, as they are natural sources of rare elements, including niobium, tantalum, titanium and the rare earths.

Crystals of pyrochlore on aegirine with microcline from Malawi.

Betafite

57

Formula $(Ca,Na,U)_2(Ti,Nb,TA)_2(O,OH)_7$
System Cubic
Habit Forms octahedral crystals; red, greenish brown, dark brown to black if high in uranium, with adamantine luster turning to earthy through alteration.
Environment Found in intrusive igneous rocks, such as granite pegmatites and carbonatites.
Name and notes Discovered near Betafo, central Madagascar, after which it is named.

Betafite crystals from Madagascar.

Microlite

58

5-5.5

5.90
6.43

Formula
$(Ca,Na)_2(Ta,Nb)_2(O,OH,F)_7$
System Cubic
Habit Usually forms octahedral crystals, commonly modified by cubic and rhombododecahedral facets; pale yellow, brown, reddish orange, brownish red, yellow-green, emerald-green, black, gray and colorless, with an adamantine luster; also found in granular and massive forms.
Environment Found in granite pegmatites; also forms through the replacement of other minerals of niobium and tantalum.
Name and notes From the Greek *mikro*, "small," referring to the minute size of its crystals. First described near Chesterfield in Hampshire County, Massachusetts.

Small octahedral crystals of microlite on elbaite from Pakistan.

Pyrochlore

59

5-5.5

4.45
4.90

Formula
$(Ca,Na)_2Nb_2(O,OH,F)_7$
System Cubic
Habit Forms octahedral crystals, commonly modified by cubic and rhombododecahedral facets; blackish brown, brown, brownish red and reddish orange, with an adamantine luster.
Environment Found in intrusive igneous rocks, such as granitic pegmatites, nepheline syenites, alkaline granite and carbonatites.
Name and notes From the Greek words *pyr*, "fire," and *chloros*, "greenish yellow," a reference to the color it assumes if heated. First discovered at Fredriksvärn, Norway.

Octahedral crystals of pyrochlore from Malawi.

RUTILE GROUP

In nature, about a dozen minerals are known to belong to this group of oxides. Those illustrated here, somewhat common, belong to the tetragonal system and have a general formula of the type $M^{4+}O_2$, in which M can represent germanium, manganese, lead, silicon, tin, tellurium and titanium. Cassiterite and rutile, aside from having beautiful and distinctive crystals, constitute an important resource for industry for the extraction of tin and titanium.

Twinned crystals of rutile with calcite from the Gries Pass in the Valais canton, Switzerland.

Cassiterite

60

6-7

6.98
7.01

Formula SnO_2
System Tetragonal
Habit Forms prismatic crystals that terminate in bipyramids, commonly twinned; black, blackish brown, reddish brown, red or brownish yellow, with an adamantine to metallic luster; also forms radiating, mammillary and granular aggregates or compact masses.
Environment Found in high-temperature hydrothermal veins; also present in granitic pegmatites and some metamorphic rocks; cassiterite can form large deposits of alluvial detritus.

Name and notes From the Greek *kassiteros*, "tin," in reference to its composition.

Large bipyramidal crystal of cassiterite from the People's Republic of China.

Plattnerite

61

5.5 | 9.56

Formula PbO_2
System Tetragonal
Habit Forms curious, usually slender, acicular crystals with acute terminations, commonly twinned; black or blackish brown, with a metallic luster.
Environment Forms in low-temperature hydrothermal veins through the alteration or oxidation of galena and other minerals of lead.
Name and notes Named after Kestner Plattner (1800–58), Professor of Metallurgy at the University of Freiberg, Germany. Discovered near Leadhills, Scotland.

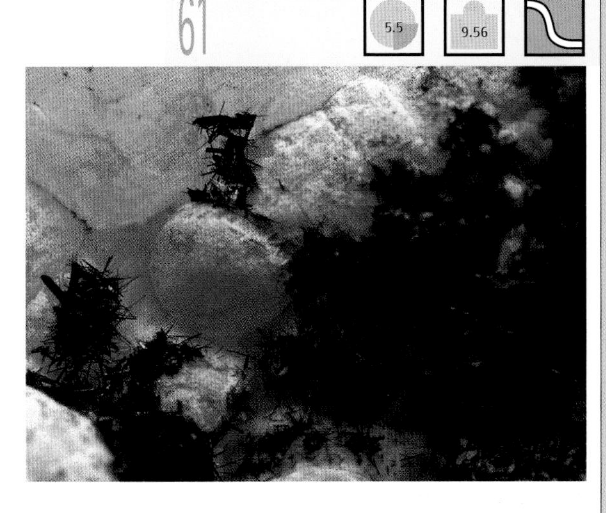

Tiny acicular crystals of plattnerite from the Dossena mine near Bergamo, Italy.

Pyrolusite

62

6–6.5 | 5.04 5.08

Formula MnO_2
System Tetragonal
Habit Forms fibrous-radiating aggregates, columnar, mammillary, dendritic or massive, with a finely granular appearance; far rarer are prismatic crystals; black or iron-gray, with a metallic luster.
Environment Forms under oxidizing conditions, in the superficial area of deposits rich in minerals of manganese.
Name and notes From the Greek *pyr*, "fire," and *louxo*, "to wash," since it was used as an additive in the manufacture of glass to eliminate green tints produced by impurities.

Aggregates of dendritic crystals from Hungary.

Rutile

63

6-6.5 | 4.21 4.25

Formula TiO_2
System Tetragonal
Habit Forms beautiful prismatic crystals, commonly long and striated, terminating in bipyramids, very commonly twinned; reddish brown, red, yellowish, black or bluish, with an adamantine or metallic luster.
Environment Occurs as an accessory mineral in intrusive igneous rocks; also common in metamorphic rocks, such as gneisses, schists and limestones; can form large alluvial deposits.
Name and notes From the Latin *rutilus*, "reddish," an allusion to the color of the mineral under a ray of light.

Above: Twinned crystals of rutile from the Formazza Valley, Piedmont, Italy.

Below: Acicular crystals of rutile from the Formazza Valley.

SPINEL GROUP

This group numbers more than 20 oxides; those illustrated here are relatively common, but the others range from the somewhat rare to the extremely rare. They have cubic symmetry and a general formula of the type AB_2O_4 (A = Co, Cu, Mg, Mn, Ni, Ti; B = Al, Cr, Fe, Mg, Mn, Ti, V). They have very similar morphological characteristics and in general form octahedral crystals. Magnetite is an important iron mineral with many industrial uses.

Specimen of spinel from Mount Somma-Vesuvius, Italy.

Chromite

64

5.5

4.50
4.80

Formula $Fe^{2+}Cr_2O_4$
System Cubic
Habit Usually forms finely granular masses; more rarely octahedral crystals modified by cubic and rhombododecahedral faces; black or blackish brown, with a metallic luster.
Environment Forms in basic and ultrabasic rocks of layered intrusive complexes and in peridotite; a somewhat common mineral in meteorites and basaltic rocks from the moon.
Name and notes The name refers to the significant quantity of chromium in the mineral's chemical composition.

Crystals with rounded edges from the Cannobina Valley, Piedmont, Italy.

Franklinite

65

6 | 5.05 5.22

Formula $ZnFe^{3+}{}_2O_4$
System Cubic
Habit Forms granular or compact masses, and also typically forms octahedral crystals; black, reddish brown or red, with a metallic luster.
Environment Occurs in deposits formed by the metamorphism of carbonate-rich rocks, associated with other minerals of iron, zinc and manganese.
Name and notes Named after Franklin, New Jersey, where it was found. Once an important ore of zinc, today, collectors seek out the crystallized specimens, particularly those from Franklin.

Crystal of franklinite in calcite from Franklin, New Jersey.

Gahnite

66

7.5-8 | 4.38 4.60

Formula $ZnAl_2O_4$
System Cubic
Habit Forms typical octahedral crystals, more rarely rhombododecahedral; dark green, bluish green, blue, yellow or brown, with a vitreous luster; also occurs in granular and compact masses.
Environment Accessory mineral in granites and granite pegmatites; also occurs in metamorphic rocks associated with sulfides.
Name and notes Named after the Swedish chemist and mineralogist Johann Gahn (1745–1818), who first recognized this mineral in 1807.

Dark blue octahedral crystals of gahnite from Tiriolo, Calabria, Italy.

Magnetite

67

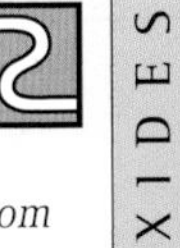

Dodecahedral crystals from a mine at Traversella, Ivrea, Italy.

Formula $Fe^{2+}Fe^{3+}_2O_4$
System Cubic
Habit Forms octahedral and in some cases dodecahedral crystals, striated, and, more rarely, cubic; black or gray, with a metallic luster; common in granular, compact or skeletal masses.
Environment Accessory mineral in igneous and metamorphic rocks; can form large deposits in the ancient sedimentary rocks called banded iron formations; also forms large deposits of alluvial detritus.
Name and notes A mineral of ancient fame, and the source of many legends because of its magnetic properties. Thales (625–547 B.C.) named it *magnete* because it had first been found in the district of Magnesia in ancient Thessaly.

Spinel

68

7.5-8 3.60 4.10

Formula $MgAl_2O_4$
System Cubic
Habit Commonly forms octahedral crystals modified by rhombododecahedral and cubic facets, often twinned; black, brown, red, orange, yellow, green, blue, violet, even colorless, with a vitreous luster; also found in granular, sometimes rounded aggregates.
Environment Accessory mineral of igneous rocks like basalts, kimberlites and peridotites; also present in metamorphic rocks like marble and schist.
Name and notes From the Latin *spina*, "thorn," alluding to its pointed crystals. The transparent colored specimens are cut to produce valuable gemstones.

Octahedral crystal of spinel from Ceylon.

Aeschynite-(Y)

69

5-6

4.82
4.93

Formula (Y,Ca,Fe,Th)(Ti,Nb)$_2$(O,OH)$_6$
System Orthorhombic
Habit Forms tabular or prismatic crystals; pale yellow, orange-yellow, reddish brown or dark brown, with a metallic or resinous luster.
Environment Occurs in granites, granitic pegmatites and carbonatites; can form alluvial deposits.
Name and notes The name is from the Greek *æschyne*, "shame"; when the mineral was discovered about a century ago, chemists were unable to separate some of its constituent elements.

Crystal in granitic pegmatites from the Vigezzo Valley, Piedmont, Italy.

Anatase

70

5.5-6

3.79
3.97

Formula TiO_2
System Tetragonal
Habit Forms pointed bipyramidal or tabular crystals, and prismatic crystals are less common. The crystals show complex forms and are black, brown, yellow, red, orange, green or blue, with a metallic or adamantine luster.
Environment Found in hydrothermal veins and inside metamorphic rocks; also found in igneous rocks such as granites, granitic pegmatites and carbonatites.
Name and notes From the Greek *anatasis*, "elongate," because it can form very long bipyramidal crystals. Much prized by collectors, who value the form and luster of its crystals.

Bipyramidal crystal from St. Christophe d'Oisan, France.

Bixbyite

71

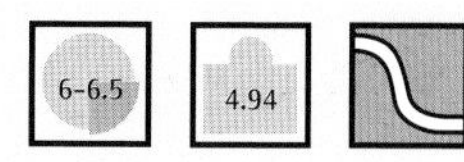

Formula $(Mn,Fe)_2O_3$
System Cubic
Habit A fairly rare mineral, it forms cubic crystals in some cases modified by octahedral faces; black, with a brilliant metallic luster.
Environment Forms in cavities of effusive igneous rocks, particularly alkaline rhyolites; also present in deposits rich in minerals of manganese.
Name and notes Named after the American mineral collector Maynard Bixby (1853–1935), author of books on the minerals of Utah.

Crystal of bixbyite showing its brilliant metallic luster, associated with an amber topaz crystal from Utah.

Brookite

72

5.5-6 | 4.08 4.18

Formula TiO^2
System Orthorhombic
Habit Forms beautiful elongate and striated tabular crystals; brown, brown-yellow, yellow, brownish red or brownish green, with metallic or adamantine luster.
Environment Forms in hydrothermal veins and in metamorphic rocks.
Name and notes Named after the British mineralogist Henry Brooke (1771–1857). The mineral was first described in the locality of Snowden, Wales. Brookite, rutile and anatase are polymorphs: they have the same chemical composition but differ in their crystalline structure.

Tabular crystal of brookite from Valais, Switzerland.

Chrysoberyl

73

Formula $BeAl_2O_4$
System Orthorhombic
Habit Forms tabular or prismatic crystals, striated; green, yellow or brown, with a vitreous luster; also found as heart-shaped penetration twins.
Environment Forms in granitic pegmatites.
Name and notes From the Greek *chrysos*, "golden yellow," and *beryllos*, "beryl," since in ancient times it was considered a yellow variety of beryl. Transparent colored varieties are cut to produce gemstones of great value.

Cyclic twinned crystal from the Ural Mountains, Russia.

Cuprite

74

Formula Cu_2O
System Cubic
Habit Forms octahedral, more rarely dodecahedral, crystals or aggregates of very thin elongate crystals with a reticulate habit; ruby-red or violet-red, with an adamantine luster.
Environment Forms in hydrothermal deposits and superficial deposits rich in copper minerals.
Name and notes From the Latin *cuprum*, "copper," in reference to its high content of that element.

Octahedral crystals of cuprite from the Democratic Republic of Congo.

Euxenite-(Y)

75

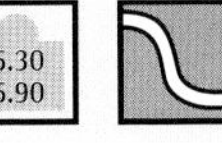

Formula (Y,Ca,REE,U,Th) $(Nb,Ta)_2(O,OH)_6$
System Orthorhombic

Habit Flattened prismatic crystals form parallel or radiating aggregates; black or blackish brown, with metallic or resinous luster; also present in compact, glassy masses if it contains high percentages of uranium.
Environment Accessory of granites and granitic pegmatites.
Name and notes From the Greek *euxenos*, "hospitable," in reference to the numerous rare chemicals that this mineral contains. First identified in the locality of Jölster, near Söndfjord, Norway.

Crystals in pegmatites from the Vigezzo Valley, Piedmont, Italy.

Ferberite

76

5-5.5 | 7.30 7.54

Formula $FeWO_4$
System Monoclinic
Habit Usually forms elongate and prismatic or tabular and striated crystals; black or blackish brown, with an adamantine or almost metallic luster.
Environment Occurs in hydrothermal quartz veins associated mostly with sulfides, carbonates and fluorite.
Name and notes Named after German mineral collector Rudolph Ferber (1805–75). First described in the Sierra Almagrera, Spain. Forms a series with hübnerite.
Tabular crystals on quartz from Peru.

Ferrocolumbite and Manganocolumbite

77

6

5.20
6.65

Formulas $Fe(Nb,Ta)_2O_6$ and $Mn(Nb,Ta)_2O_6$
System Orthorhombic
Habit These minerals, which have similar characteristics, form prismatic to tabular crystals, somewhat flattened, commonly joined in parallel aggregates; black, blackish brown, reddish brown where iron is dominant, and red where manganese is dominant, with metallic luster.
Environment Accessory in granite pegmatites, more rarely of carbonatites; can form alluvial deposits of great industrial importance.
Name and notes Columbite is from the chemical element *columbium*, named after Christopher Columbus (and later renamed niobium). Ferrocolumbite and manganocolumbite form a series, having the same crystalline structure but differing in the proportion of iron and manganese.

Above left and below: Specimens from granitic pegmatites from the region of Mahaiza, Madagascar.

Manganotantalite

6-6.5

6.65
7.95

Formula $Mn(Ta,Nb)_2O_6$
System Orthorhombic
Habit Forms prismatic or tabular crystals, somewhat flattened, commonly joined in parallel aggregates; black, reddish brown or red, with a metallic or vitreous luster.
Environment Accessory mineral in granite pegmatites; can form alluvial deposits of great importance to industry.
Name and notes The name tantalite is from the chemical element tantalum, named in turn after the mythical Tantalus, in reference to the great difficulties experienced by the mineral's discoverers when they tried to dissolve it in acids to analyze it. Manganotantalite's iron-dominant analogue, ferrotantalite (in which the iron content predominate over the manganese), exists in nature, but it is very rare.

Above: Crystal of manganotantalite from Afghanistan.
Left: Crystals on elbaite from Afghanistan.

Hübnerite

79

Formula $MnWO_4$
System Monoclinic
Habit Forms prismatic crystals, tabular and striated; blackish brown, brownish red to yellow, with an adamantine or metallic luster.
Environment Found in hydrothermal veins formed at high temperature, associated with cassiterite, apatite, arsenopyrite, schorl, topaz and fluorite; also found in granitic pegmatites.
Name and notes Named after the German chemist and metallurgist Adolph Hübner. Forms an isomorphic series with ferberite, meaning it has the same crystalline structure but the manganese content predominates over the iron.

Prismatic crystals of hübnerite on quartz from Peru.

Ilmenite

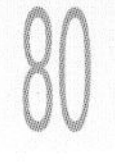

5.6 4.72

Formula $FeTiO^3$
System Trigonal
Habit Forms thick tabular or rhombohedral crystals; black or gray, with a metallic luster; also occurs in skeletal or massive granular habit.
Environment Accessory of many intrusive igneous rocks such as granites, gabbros, kimberlites, granitic pegmatites and carbonatites; present in metamorphic rocks; can form large alluvial deposits of great economic interest.
Name and notes Named after Lake Ilmen, in the Ural Mountains of Russia, where the mineral was discovered.

Crystal from quartzite veins in the Malenco Valley, Lombardy, Italy.

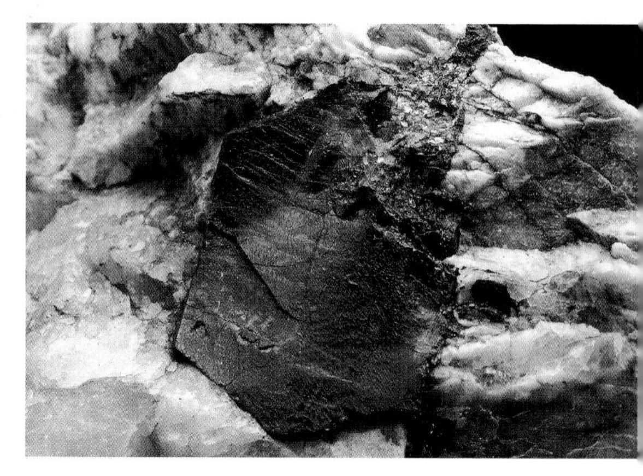

Opal

6.5

2.10

Formula $SiO_2 \cdot nH_2O$
System None (amorphous)
Habit Presents a disordered crystalline structure, not forming distinct crystals but rather globular, stalactitic and mammillary masses; white, yellow, red, orange, green, brown, blue, black or colorless, with a vitreous luster.
Environment Forms through the circulation of silica-rich water in sedimentary rocks, also replacing fossils; present in volcanic rocks, intrusive igneous rocks and pegmatites as fillings of cavities and fractures.
Name and notes A mineral known since antiquity, its name is from the Greek *opalios* and the Sanskrit *upala*, "precious stone." Some colored and transparent varieties are cut to produce gemstones of great value.

Above: Specimen of opal from Mexico.
Below, left and right: Examples of "noble" opal from Australia.

Quartz

7

2.65

Formula SiO_2
System Trigonal
Habit Extremely widespread, with hundreds of different crystalline habits, including prismatic and rhombohedral; twins are common, as are somewhat complex crystals, including those of pseudocubic habit; colorless, white, pink, reddish, yellow, green, blue, violet or brown, with a vitreous luster; common in groups of parallel crystals and in microcrystalline aggregates.
Environment Widespread mineral found in hydrothermal veins, in granite and granitic pegmatites; also very common in metamorphic and sedimentary rocks.
Name and notes Probably from the German *Quarz* or the Slavic *kwardy*. Until the end of the 18th century, quartz was commonly called rock crystal, but in his *De Re Metallica* of 1556, German mineralogist and metallurgist Georgius Agricola (1494–1555) already used the word *querz*.

Left: Crystals of smoky quartz from Switzerland.

Top: Specimen of quartz from Arkansas.
Above: Quartz with hematite from the People's Republic of China.

Thorianite

83

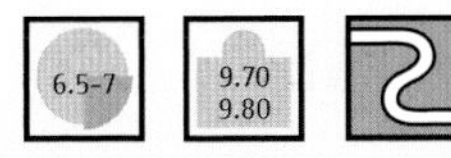

Formula ThO_2
System Cubic
Habit Typically forms cubic crystals, in some cases modified by octahedral faces; black, brownish black to dark gray, with a metallic luster, resinous where the mineral has been altered; also present in somewhat rounded granular aggregates.
Environment Accessory mineral in granitic pegmatites and carbonatites; more rarely present in metamorphic rocks.
Name and notes The name reflects its high level of thorium, an element named for Thor, the Norse god of thunder. The mineral is radioactive and has industrial uses.

Thorianite penetration-twin crystal from Quebec, Canada.

Uraninite

84

5-6
10.63
10.95

Formula UO_2
System Cubic
Habit Forms cubic, cubo-octahedral or rhombododecahedral crystals; steel-black, brownish black to greenish black, with metallic luster; also occurs in massive and mammillary forms.
Environment Occurs in granite, in syenites and pegmatites, present in hydrothermal deposits associated with minerals of cobalt, bismuth and arsenic, and in volcano-sedimentary rocks.
Name and notes The name refers to its high content of uranium, an element whose name is derived from the planet Uranus. Forms an isomorphic series with thorianite (i.e., uranium can substitute for thorium in the same structure). Radioactive and of great strategic interest, particularly in the production of nuclear energy.

Specimen of uraninite from Wölsendorf, Bavaria.

Brucite

85

Formula Mg(OH)^2
System Trigonal
Habit Common in foliated, mammillary, fibrous and massive aggregates, more rarely as distinct crystals; tabular with a pseudohexagonal outline; white, green, blue or gray, with a pearly luster.
Environment Found in hydrothermal veins in marble and chloritic schist; a typical product of alteration of periclase, an oxide of magnesium unstable in natural conditions.
Name and notes Named after Archibald Bruce (1777–1818), the Yale University physics professor who first described it.

Below and bottom: Groups of foliate brucite crystals from the Astico Valley, near Vicenza, Italy.

Becquerelite

86

2.5 | 5.09 5.20

Formula
$Ca(UO_2)_6O_4(OH)_6\_8H_2O$
System Orthorhombic
Habit Forms beautiful tabular or prismatic crystals, striated with pseudohexagonal outline; amber-yellow, golden yellow, orange-yellow or yellowish brown, with an adamantine luster.
Environment Product of alteration in zones of secondary oxidation in deposits rich in uranium; more rarely found in granitic pegmatites.
Name and notes Named after French physicist Antoine Henri Becquerel (1852–1908), who discovered radioactivity in uranium and was awarded the 1903 Nobel Prize in Physics (with the Curies). Radioactive and highly valued by collectors.

Prismatic becquerelite crystals with uranophane from Shaba, Democratic Republic of Congo.

Diaspore

Formula AlOOH
System Orthorhombic
Habit Usually forms stalactitic aggregates of laminar or foliated appearance, also very rare as distinct crystals, flattened and somewhat elongate; white, gray, greenish gray, brown, yellow, pink or colorless, with an adamantine or vitreous luster.
Environment Typical product of alteration in bauxite deposits, rich in silicates of aluminum, under conditions of tropical weathering; also present in hydrothermal veins, and more rarely found in alkaline pegmatites.
Name and notes From the Greek *diasphrein*, "breakable," recalling its characteristic to crack when heated.

Diaspore from the Emery mine, Massachusetts.

Goethite

88

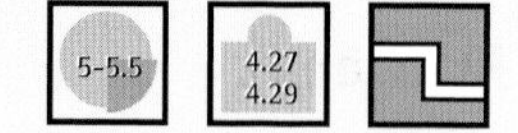

Formula FeOOH
System Orthorhombic
Habit Forms prismatic crystals, tabular and striated; dark brown to reddish brown with metallic luster, earthy if the mineral has been altered; very widespread in stalactitic form and as mammillary or fibrous-radiating aggregates.
Environment Typical product of superficial alteration in deposits rich in iron minerals.
Name and notes Named after the celebrated German poet, dramatist, novelist, scientist and mineralogist Johann Wolfgang von Goethe (1749–1832).

Below: Specimen of goethite from Saxony, Germany.

Bottom: Mammillary aggregates of goethite from Rio Marina, Elba, Italy.

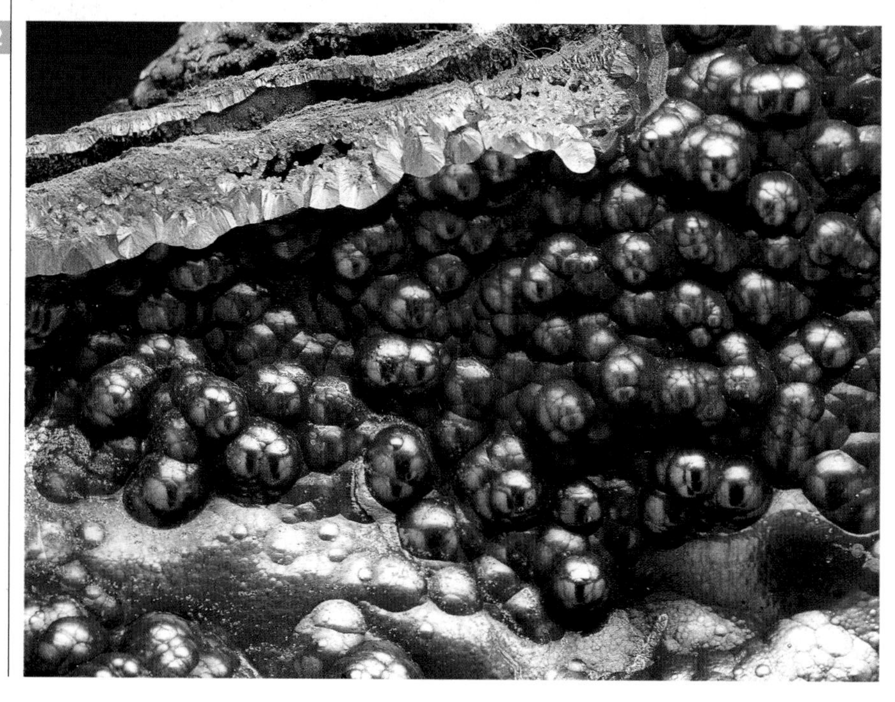

Manganite

4

4.29
4.34

Formula MnO(OH)
System Monoclinic
Habit Forms groups of prismatic crystals, parallel, striated and with flat terminations; steel-gray or black, with a metallic luster; also found in fibrous, granular and compact aggregates.
Environment Occurs in hydrothermal formations where it can form large superficial deposits, associated with other minerals of manganese; a characteristic product of alteration of deposits of manganese.
Name and notes The name reflects its high content of manganese. It was first identified in the locality of Ilfeld in the Harz region of Germany.

Top and left: Groups of striated prismatic crystals of manganite from Ilfeld, Germany.

CARBONATES

Class 5

ARAGONITE GROUP

The minerals in this group have an orthorhombic symmetry and the general formula ACO_3, in which A can represent barium, calcium, lead and strontium. Commonly found in twinned crystals that characteristically simulate hexagonal symmetry; except for witherite, these are widespread minerals, aragonite in particular. Being carbonates they are easily affected by acids.

Aragonite crystals with sulfur from Sicily.

Aragonite

90

3.5-4

2.94
2.95

Formula $CaCO_3$
System Orthorhombic
Habit Forms prismatic crystals, tabular or acicular; white, gray, green, pink-red, violet or colorless, with a vitreous luster; common as twinned crystals with a pseudohexagonal section and in radiating or columnar aggregates.
Environment A mineral characteristic of sedimentary rocks, also occurs in metamorphic and igneous rocks, where it forms through the circulation of aqueous solutions; it is the principal component of stalactites and stalagmites.

Aragonite from Australia.

Name and notes Named after the region of Aragon in Spain, where the mineral was first found. Aragonite is the high-pressure polymorph of $CaCo_3$, having the same chemical composition but a different crystalline structure than calcite.

Cerussite

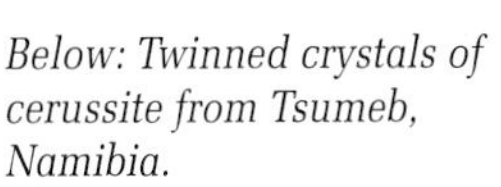

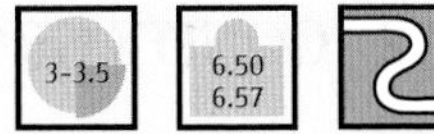

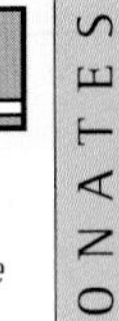

Formula $PbCO_3$
System Orthorhombic
Habit Forms tabular crystals, bipyramidal, pseudohexagonal or acicular, frequently striated and twinned; white, gray, brown and colorless, with an adamantine luster.
Environment A typical product of the alteration of galena; found in the superficial oxidized zones of deposits rich in minerals of lead, where it is associated with numerous other alteration minerals.
Name and notes From the Latin *cerussa*, "ceruse," or white lead as a pigment, the name by which this mineral was known in antiquity.

Below: Twinned crystals of cerussite from Tsumeb, Namibia.

Bottom: Crystals of cerussite from Morocco.

Strontianite

92

3.5 3.70 3.76

Formula $SrCO_3$
System Orthorhombic
Habit Forms prismatic crystals, striated, also twinned in pseudohexagonal section or acicular; gray, yellow, pale green, yellowish brown or reddish, with a vitreous luster; also common in massive, columnar and fibrous forms.
Environment Occurs in hydrothermal formations, associated with carbonate-rich and metamorphic rocks of sedimentary origin, notably marble.
Name and notes Named after Strontian, Argyll, Scotland, where the mineral was first recognized in 1790.

Group of acicular crystals of strontianite from England.

Witherite

93

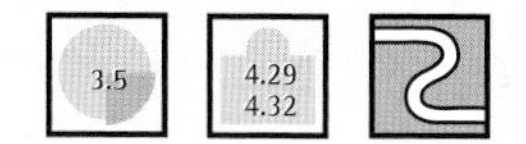

Formula $BaCO_3$
System Orthorhombic
Habit Forms crystals that are invariably twinned, horizontally striated, creating pseudohexagonal bipyramids; white, yellow or gray, with a vitreous luster; also occurs as columnar, fibrous, granular or compact aggregates.
Environment Mineral of hydrothermal veins, associated with numerous other low-temperature minerals, in particular sulfides, carbonates and sulfates.
Name and notes Named after British physicist and naturalist William Withering (1741–99), the first to describe this mineral. Discovered near Alston Moor, England.

Characteristic twinned crystal of witherite from Alston Moor, England.

CALCITE GROUP

About a dozen carbonates belong to this group; those illustrated here are very widespread, in particular calcite, which is an important component of many carbonate-rich rocks.

These minerals have trigonal symmetry and the general formula of the type A[CO_3], in which A can represent calcium, cadmium, cobalt, iron, magnesium, manganese, nickel and zinc. The minerals in this group cleave perfectly and dissolve readily in acid.

Contact twin of calcite from the People's Republic of China.

Calcite

94

3

2.71
2.94

Formula $CaCO_3$
System Trigonal
Habit Extremely widespread; hundreds of different crystalline habits have been described, including rhombohedral, scalenohedral, prismatic hexagonal or complex combinations of forms, and twinned crystals are common; also occurs as lamellar, columnar, stalactitic, fibrous and granular aggregates; in all tonalities of colors, colorless or white, with vitreous or dull luster.
Environment The principal component of many sedimentary rocks and stalactites and stalagmites; found in many environments, including effusive igneous rocks and metamorphic rocks.
Name and notes From the Greek *chalx*, "lime." Colorless calcite crystals are used to make polarizing optical lenses.

Prismatic hexagonal crystals of calcite from the United States.

Magnesite

95

3.5–4.5 | 2.98

Formula $MgCO_3$
System Trigonal
Habit Forms granular, lamellar and fibrous masses; crystals are rare, usually rhombohedral or prismatic; white, gray, yellow and brown in varieties rich in iron, with a vitreous luster.
Environment Occurs in deposits, sometimes large, associated with metamorphic rocks like serpentinites; constituent of effusive igneous rocks, such as carbonatites; also occurs in sedimentary rocks.
Name and notes From the Greek *magnesia lithos*, "magnesium stone," the rough mineral presumably extracted near Magnesia, a coastal province of ancient Thessaly.

Magnesite specimen from Italy.

Rhodochrosite

96

3.5–4 | 3.70

Formula $MnCO_3$
System Trigonal
Habit Common in massive and granular forms, or in globular, fibrous and columnar aggregates; less common are crystals, rhombohedral, scalenohedral and prismatic; pink, red and reddish brown, with a vitreous luster.
Environment Occurs in hydrothermal veins associated with minerals of silver, lead, copper and zinc; also occurs in deposits formed through the metamorphism of carbonate-rich rocks.
Name and notes From the Greek words *rhodon*, "rose," and *chros*, "color," referring to its characteristic color.

Splendid specimen of rhodochrosite crystals from Colorado.

Siderite

97

Formula $FeCO_3$
System Trigonal
Habit Forms massive, granular, botryoidal and fibrous aggregates; also found in tabular, prismatic and rhombohedral crystals, often with curved faces; yellow-brown, red-brown, brown and dark brown from surface oxidation, with a vitreous luster where fresh.
Environment Generally in sedimentary deposits; also widespread in hydrothermal veins associated with minerals of iron, manganese and lead; present in intrusive igneous rocks such as granites and pegmatites.
Name and notes From the Greek *sideros*, "iron," referring to its composition.

Curving crystals from Piedmont, Italy.

Smithsonite

98

Specimen of smithsonite from Arizona.

Formula $ZnCO_3$
System Trigonal
Habit Forms rhombohedral crystals, commonly with curved faces that can make distinctive botryoidal aggregates; white, gray, greenish, brown, yellow, pink, bluish green or blue, with a vitreous luster; very common in stalactitic, botryoidal, coarse and earthy aggregates.
Environment Common in surficial or near-surface deposits rich in minerals of zinc, and associated with limestone and dolomites, in some cases forming large deposits.
Name and notes Named for British chemist and mineralogist James Smithson (1754–1829), founder of the Smithsonian Institution in Washington, D.C.

DOLOMITE GROUP

Five minerals belong to this group, all with trigonal symmetry and a general formula of the type $AB(CO_3)_2$ in which A = Ba, Ca and B = Fe, Mn, Zn. Dolomite and ankerite, quite common on our planet, are fundamental components of many carbonate rocks, and they can form a partial or complete isomorphic series.

Twinned crystals of dolomite with quartz from Traversella, Piedmont, Italy.

Ankerite

99

Formula $CaFe(CO_3)_2$
System Trigonal
Habit Forms rhombohedral crystals and granular aggregates; brown or dark brown through weathering, with a vitreous luster where fresh.
Environment Principal component of certain sedimentary and igneous rocks, also occurs in hydrothermal veins and in metamorphic rocks.
Name and notes Named after Austrian mineralogist Mathias Joseph Anker (1771–1843).

Specimen of ankerite from Schilpario, Lombardy, Italy.

Dolomite

Formula $CaMg(CO_3)_2$
System Trigonal
Habit Commonly forms rhombohedral crystals, often twinned; white, gray, pale green, brown or colorless, with a vitreous luster; also occurs in massive and granular forms.
Environment Principal component of many sedimentary rocks, common mineral in hydrothermal veins, principal component of effusive igneous rocks, such as carbonatites, and of some metamorphic rocks, in particular marble.
Name and notes Named after French geologist and mineralogist Déodat de Dolomieux (1750–1801), who first described the mineral in 1791.

Twinned crystals of dolomite from marble quarries in Crevoladossola, Piedmont, Italy.

Kutnohorite

Formula $CaMn(CO_3)_2$
System Trigonal
Habit Somewhat rare; in general forms fibrous and granular aggregates; pink or white, with a vitreous luster.
Environment Occurs in deposits formed by the metamorphism of carbonate-rich rocks associated with other minerals of manganese.
Name and notes Named after Kutná Hora, Bohemia, Czech Republic, where it was first described.

Specimen of kutnohorite from South Africa.

Artinite

102

2.5

2.02
2.04

Formula
$Mg_2(OH)_2 (CO_3)\cdot 3H_2O$
System Monoclinic
Habit Forms radiating and spherical aggregates of acicular, elongate, and fibrous crystals; white, with a silky luster.
Environment Occurs in hydrothermal veins in association with ultramafic metamorphic rocks, in particular serpentinites.
Name and notes Named after Ettore Artini (1866–1928), Professor of Mineralogy and director of the Museum of Natural History in Milan. It was first described in the locality of Campo Franscia in the Malenco Valley, Lombardy, Italy.

Artinite crystals from the Varaita Valley, Piedmont, Italy.

Aurichalcite

103

1-2

4.00

Formula $(Zn,Cu)_5(OH)_6(CO_3)_2$
System Monoclinic
Habit Forms elongate acicular crystals found in globular, feathery and fibrous aggregates; green, blue and sky-blue, with a silky luster.
Environment Somewhat common in surficial or near-surface deposits rich in minerals of copper and zinc; forms in sedimentary rocks enriched in carbonate.
Name and notes From the Latin, *aurichalcum*, "copper yellow," which is curious since in reality the mineral is blue.

Groups of acicular crystals from the Murvonis mine, Sardinia, Italy.

Azurite

Formula $Cu_3(OH)_2(CO_3)_2$
System Monoclinic
Habit Occurs in a great variety of habits, forming beautiful tabular crystals, prismatic and commonly striated; azure-blue or dark blue in color, with a vitreous luster; also common in stalactitic and globular aggregates.
Environment Secondary mineral that forms in surficial or near-surface areas of copper deposits, where hydrothermal fluids rich in copper salts circulate; also forms in association with malachite and other copper minerals.

Name and notes From the Persian word *lazhuward,* "sky-blue."

Interesting specimen of azurite from Tsumeb, Namibia.

Hydrozincite

105

Formula $Zn_5(OH)_6(CO_3)_2$
System Monoclinic
Habit Common in massive, earthy and fibrous forms, as radiating or stalactitic masses; white, gray, pale yellow or pinkish brown, with a dull luster.
Environment Secondary mineral, characteristic of deposits rich in minerals of zinc, also forms in carbonate-rich sedimentary rocks through the alteration of sphalerite.
Name and notes The name reflects its chemical composition.

Fibrous crystals of hydrozincite from Mexico.

Malachite

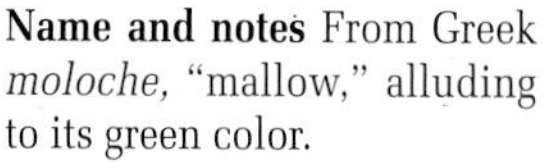

3.5-4 4.05

Formula $Cu_2(OH)_2(CO_3)$
System Monoclinic
Habit Forms crystals, somewhat rare, generally prismatic or joined in aggregates of acicular crystals, fibrous and radiating; pale green to emerald-green, with a luster that varies from adamantine (crystals) to dull (massive); common in stalactitic form and as aggregates with a concentric structure.
Environment Secondary mineral that forms in surficial parts of copper deposits through the circulation of water rich in dissolved copper in association with other copper minerals.
Name and notes From Greek *moloche,* "mallow," alluding to its green color.

Beautiful specimen of malachite from the Democratic Republic of Congo.

Parisite-(Ce)

107

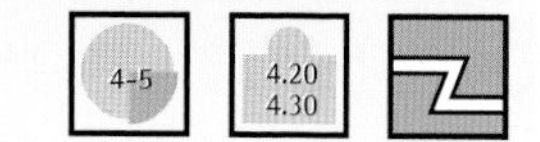

Formula $Ca(Ce,La)_2F_2(CO_3)_3$
System Trigonal
Habit Forms prismatic crystals that are hexagonal, usually terminating in bipyramids or pinacoids; brown, brownish yellow or orange-brown, with vitreous or resinous luster.
Environment Occurs in alkaline granite pegmatites and in hydrothermal veins, associated with calcite and beryl.
Name and notes Named for Jean Paris, owner of the mine at Muzo, Boyaca, Colombia, where the mineral was discovered.

Prismatic hexagonal crystal of parisite-(Ce) on aegirine from the alkaline pegmatites of Mount Malosa, Malawi.

Phosgenite

2-3

6.13

Formula $Pb_2Cl_2(CO)_3$
System Tetragonal
Habit Forms prismatic crystals, in some cases tiny, other times long and tabular; yellowish, amber, brown, pink, grayish, white or colorless, with an adamantine luster.
Environment Secondary mineral that forms in metalliferous deposits rich in minerals of lead, in particular through the alteration of galena, through the circulation of hydrothermal fluids with high content of dissolved salts.
Name and notes Named after the chemical term *phosgene,* the elements of which (carbon, chlorine and oxygen) are present in this mineral. Highly valued by collectors for its form and the brilliance of its crystals.

Top and above: Specimens of phosgenite from Monteponi, Sardinia, Italy.

BORATES

Class 6

Behierite

109

Formula $(Ta,Nb)BO_4$
System Tetragonal
Habit Forms bipyramidal crystals; pale brown, pinkish brown or dark brown, with an adamantine luster.
Environment Very rare, occurs in granite pegmatites rich in cesium, where it is associated with rhodizite.
Name and notes Named after Jean Behier (1903–63), director of the research laboratory of the Geological Service of Madagascar. The mineral was first identified in the locality of Manjaka, Madagascar.

Characteristic isolated bipyramidal crystals of behierite from Tetetsantsio, Madagascar.

Boracite

110

Formula $(Mg,Fe)_3ClB_7O_{13}$
System Orthorhombic
Habit Forms pseudocubic crystals; white, gray or green, with a vitreous luster.
Environment Common in sedimentary evaporite deposits, associated with numerous other minerals of boron.
Name and notes Named for its high content of boron.

Beautiful rhombohedral crystal of boracite from Germany.

Canavesite

111

n.d. 1.81

Formula $Mg_2(CO_3)(HBO_3)\cdot 5H_2O$
System Monoclinic
Habit Very rare mineral, forms thin acicular crystals joined in radiating aggregates; white, with a vitreous luster.
Environment Late mineral, found in thin fractures of metamorphic skarn containing magnetite and other minerals of boron.
Name and notes Named after the Canavese region of the province of Turin, Italy, where it was found in the Volagera level of the Brosso mine.

Typical cluster of acicular crystals of canavesite from the Brosso mine, Piedmont, Italy.

Colemanite

112

4.5 2.42

Formula $CaB_3O_4(OH)_3$_H_2O
System Monoclinic
Habit Common in massive and granular forms; also found in distinct crystals of complex habit; colorless, white or gray, with a vitreous luster.
Environment Occurs in sedimentary evaporite deposits, associated with other minerals of boron.
Name and notes Named for American industrialist William Tell Coleman (1824–93), mine owner and a founder of the California borax industry. The mineral was first identified at Furnace Creek, Death Valley, California.

Crystals of colemanite from Turkey.

Hambergite

113

7-7.5 2.36

Formula $Be_2BO_3(OH)$
System Orthorhombic
Habit Forms prismatic to tabular crystals; colorless, white or pink, with a vitreous luster.
Environment Somewhat rare mineral, found as accessory to various granitic pegmatites.
Name and notes Named for Swedish mineralogist Axel Hamberg (1863–1933), who first called attention to the study of this mineral, discovered near Helgaråen, Norway.

Hambergite crystal from Madagascar.

Jeremejevite

7-7.5 7.81 7.91

Formula $Al_6(BO_3)_5(F,OH)_3$
System Hexagonal
Habit Forms characteristic prismatic hexagonal crystals; blue, sky-blue, brownish yellow or colorless, with a vitreous luster.
Environment Rare accessory mineral in alkaline igneous rocks, particularly granite pegmatites.
Name and notes Named for Russian mining engineer Pavel Jeremejev (1830–99), who was the first to find this mineral, on Mount Soktui in Siberia.

Left and right: An isolated crystal from Erongo, Namibia.

Rhodizite

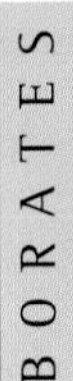

Formula
$(K,Cs)Al_4Be_4(B,Be)_{12}O_{28}$
System Cubic
Habit Forms beautiful rhombo-

Below and bottom: Two characteristically clear crystals of rhodizite from Madagascar.

dodecahedral and tetrahedral crystals; yellow, green-yellow, green, brownish yellow, white or colorless, with a vitreous or adamantine luster.

Environment Rare accessory mineral of some granitic pegmatites rich in cesium.

Name and notes From the Greek *rhodizein*, "rose colored," alluding to the pink-red color of the flame when the mineral is heated in a tube. First identified at Schaitansk, near Mursinsk, Russia.

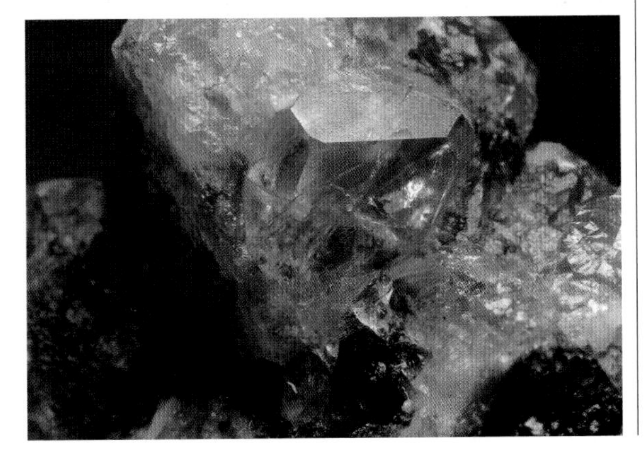

SULFATES, CHROMATES, MOLYBDATES, TUNGSTATES

BARITE GROUP

The sulfates of lead, barium and strontium that belong to this group are somewhat common minerals on our planet. They have an orthorhombic symmetry, a general formula of the type AXO_4 (in which A = Ba, Pb, Sr; and X = S), and very similar crystalline habits, but they are easy to tell apart, in particular when dealing with well-formed crystals. Such crystals are highly prized by collectors.

Barite crystals from the United States.

Anglesite

116

Formula $PbSO_4$
System Orthorhombic
Habit Found in a great variety of forms, including tabular, prismatic and bipyramidal crystals, commonly with striated faces; white, gray, yellow, green, pink, blue and colorless, with an adamantine luster.
Environment Typical product of the alteration of galena; occurs in the near-surface oxidized zone of deposits rich in minerals of lead, where it is associated with numerous other minerals of typical alteration.
Name and notes Named after the Isle of Anglesey, Wales, where it was first found.

Exceptional crystal of anglesite from Sardinia, Italy.

Barite

117

3-3.5 | 4.50

Formula $BaSO_4$
System Orthorhombic
Habit Common in tabular and prismatic crystals; white, yellow, brown, gray, reddish, pink, blue or colorless, with a vitreous luster; also common in aggregates forming rosettes and in fibrous, laminar and stalactitic forms.
Environment Typical of hydrothermal veins; also widespread in sedimentary rocks, in particular evaporites, limestones and dolomites; also present in some igneous rocks, such as carbonatites, in cavities (vesicles) in basaltic rocks and in metamorphic skarn.

Name and notes From the Greek *barys*, "heavy," alluding to its high density.

Tabular crystals of barite from Romania.

Celestine

118

3-3.5 | 3.96

Formula $SrSO_4$
System Orthorhombic
Habit Commonly forms tabular to prismatic crystals; pale blue, white, green, brown or colorless, with a vitreous luster; also common in radiating aggregates, fibrous and nodular.
Environment Forms in sedimentary rocks, particularly in evaporite formations, in limestones and through the circulation of hydrothermal fluids; also present in cavities (vesicles) in basaltic rocks.
Name and notes From the Latin *cælestis*, "celestial," referring to the sky-blue color of some varieties.

Group of prismatic crystals of celestine from the sulfur mines of Sicily, Italy.

ETTRINGITE GROUP

This group of sulfates (six are known) presents notable complexity, both in chemical and crystallochemical terms. These minerals are somewhat rare in nature; sturmanite forms beautiful yellow crystals that are very popular with collectors, whereas ettringite forms, also artificially, in industrial cements.

Typical crystals of ettringite with pseudohexagonal habit from Montalto di Castro, Italy.

Ettringite

119

Formula
$Ca_6Al_2(SO_4)_3(OH)_{12} \cdot 26H_2O$
System Hexagonal
Habit Somewhat rare, forms prismatic crystals with hexagonal contours; colorless, white, yellow and pale brown, with a vitreous luster.
Environment Found in cavities of metamorphic limestone rocks and in some effusive alkaline rocks.
Name and notes Named after the locality of Ettringen in Rhineland, Germany, where it was discovered.

Ettringite crystals from South Africa.

Sturmanite

2-2.5

1.80
1.85

Formula $Ca_6Fe^{3+}{}_2(SO_4)_2[B(OH)_4](OH)_{12} \cdot 25H_2O$

System Hexagonal

Habit Very rare mineral, forms beautiful prismatic hexagonal crystals; bright yellow or greenish yellow, with a vitreous luster.

Environment Found in cavities of metamorphic skarn associated with barite and hematite.

Name and notes Named for B. Darko Sturman (b. 1937), curator at the Royal Ontario Museum, Toronto, Canada. The mineral was first described at Kuruman in Cape Province, South Africa.

Yellow hexagonal prismatic crystals of sturmanite from the Kalahari region, South Africa.

Thaumasite

121

3.5

1.88
1.90

Formula $Ca_6Si_2(CO_3)_2(SO_4)_2(OH)_{12} \cdot 24H_2O$

System Hexagonal

Habit Occurs in fibrous massive aggregates and in groups of acicular to prismatic crystals; white.

Environment Forms in skarn rich in silicates of calcium, iron and magnesium.

Name and notes From the Greek *thaumazein*, "surprising," alluding to its unusual chemical composition. It was first described at Bjelkesgruvan, Jämptland, Sweden.

Thin acicular crystals from Braone, Camonica Valley, Italy.

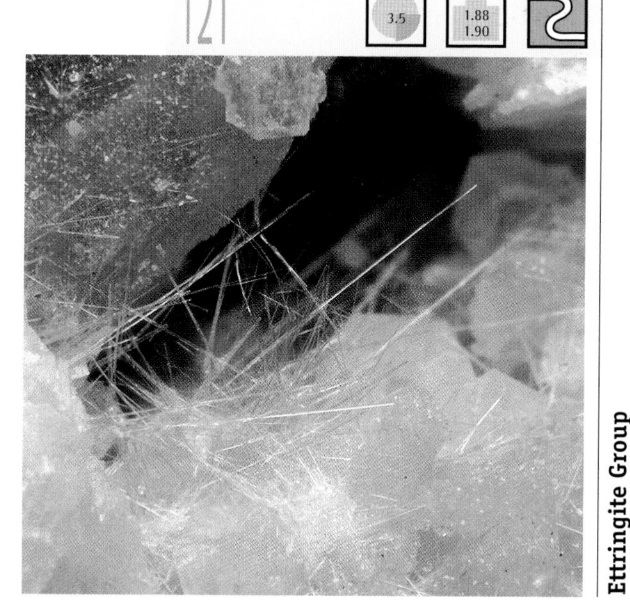

Alunite

122

3.5-4

2.59
2.90

Formula $Kal_3(SO_4)_2(OH)_6$
System Trigonal
Habit Massive, granular and fibrous forms; very rare rhombohedral crystals; white, grayish, yellowish or reddish brown, with a vitreous luster.
Environment Found in rocks that have been altered by the circulation of hydrothermal fluids rich in sulfuric acid; can form deposits of industrial interest for the extraction of aluminum.
Name and notes From the Latin *alumen*, "aluminum," a component of this mineral. First described near La Tolfa, Latium, Italy.

Specimen of alunite from El Indio, La Serena, Chile.

Anhydrite

123

Formula $CaSO_4$
System Orthorhombic
Habit Usually occurs in easily cleaved masses, more rarely as tabular crystals; colorless, white, gray, violet, pink or yellowish brown, with a vitreous luster.
Environment Principal component of certain sedimentary rocks, such as evaporites also found in alpine-type hydrothermal veins; a product of sublimation of fumaroles, it is also present in certain skarns.
Name and notes From the Greek words *an*, "without," and *hydor*, "water"; gypsum has a similar composition but contains H_2O molecules. First found near Hall, in the Austrian Tyrol.

Crystallized specimen from Campiano, Tuscany, Italy.

Brochantite

124

3.5–4 | 4.00

Formula $Cu_4(SO_4)(OH)_6$
System Monoclinic
Habit Forms prismatic, acicular or tabular crystals; emerald-green or dark green, with a vitreous luster.
Environment Found in near-surface deposits rich in minerals of copper, where it forms through the circulation of hydrothermal solutions.
Name and notes Named after André Brochant de Villiers (1772–1840), French mineralogist, professor at the École des Mines in Paris and author of a treatise on mineralogy. The mineral was first described at the Bank mines in the Ural Mountains, Russia.

Group of acicular crystals from the People's Republic of China.

Chalcanthite

125

2.5 | 2.28

Formula $CuSO_4 \cdot 5H_2O$
System Triclinic
Habit Forms prismatic and tabular crystals; deep blue, sky-blue or greenish blue, with a vitreous luster; also common in massive, granular and stalactitic forms.
Environment Forms in near-surface deposits rich in minerals of copper, where it forms through the circulation of low-temperature hydrothermal solutions.
Name and notes From the Greek *chalkos*, "copper," and *anthos*, "flower," in reference to its occurrence as an incrustation on copper minerals.

Specimen of chalcanthite from the Yorke Peninsula, Australia.

Cyanotrichite

2 | 2.74 2.95

Formula $Cu_4Al_2(SO_4)(OH)_{12} \cdot 2H_2O$
System Orthorhombic
Habit Forms aggregates of very thin acicular crystals; also found in radial and fibrous groups; sky-blue or deep blue, with a silky luster.
Environment Forms in near-surface deposits rich in minerals of copper, where it forms through the circulation of hydrothermal fluids.
Name and notes From the Greek *kyanos*, "blue," and *trichos*, "hair," alluding to its color and typical habit. It was first identified near Moldava, Romania.

Specimen of cyanotrichite from Arizona.

Gypsum

127

2 | 2.31

Formula $CaSO_4 \cdot 2H_2O$
System Monoclinic
Habit Common in tabular, prismatic or acicular crystals with striated faces, or lenticular with rounded faces; colorless, white, pale yellow or grayish, with a vitreous luster; also frequent in aggregates that are joined in rosettes, fibrous or in easily-cleaved masses.
Environment Occurs in evaporite sedimentary rocks, where it may form large deposits.
Name and notes From the Greek *gypsos*, "plaster."

Specimen of gypsum from a sulfur mine in Sicily.

Linarite

128

Formula $PbCu(SO_4)(OH)_2$
System Monoclinic
Habit An uncommon mineral found in tabular to prismatic crystals, commonly striated; deep blue, with an adamantine luster; also present in microcrystalline aggregates that form thin incrustations.
Environment Secondary mineral, forms in near-surface zones of deposits rich in lead and copper minerals through the circulation of hydrothermal solutions.
Name and notes Named after Linares, Jaen province, Spain, near where it was first found.

Typical deep blue tabular crystals of linarite from Arizona.

Voltaite

129

3-3.50

2.60
2.80

Formula
$K_2Fe^{2+}{}_5Fe^{3+}{}_3Al(SO_4)_{12} \cdot 18H_2O$
System Cubic
Habit Forms distinct cubic or octahedral crystals; greenish black, dark green or greenish gray, with a resinous luster.
Environment Typical product of sublimation in fumaroles; also present as a late mineral in galleries of mines, where it originates, together with other sulfates, from the alteration of primary sulfides.
Name and notes Named after Italian physicist Count Giuseppe Antonio Alessandro Volta (1745–1827).

Specimen of voltaite from the Tinto River, Huelva, Spain.

2.5–3

5.99

Crocoite

Formula $PbCrO_4$
System Monoclinic
Habit Usually forms prismatic crystals, often hollow; orange-red or orange-yellow, with an adamantine luster.
Environment Secondary mineral, forms in the upper, oxidized zones of deposits rich in lead that are associated with ultramafic rocks.
Name and notes From the Greek *krokon*, "saffron," referring to its color. First found near Berezov, in the Ural Mountains, Russia. Highly valued by collectors for its color and the form of its crystals.

Specimen of crocoite from Berezov, Russia.

Specimen from Tasmania.

Powellite

3.5 4.34

Formula $CaMoO_4$
System Tetragonal
Habit Typical bipyramidal crystals, tetragonal and more rarely tabular; white, brownish yellow or gray, with a vitreous luster.
Environment Forms in hydrothermal veins by the alteration of molybdenite; also found in quartz veins, associated with metamorphic rocks in skarn, more rarely in cavities of basaltic rocks associated with zeolites.
Specimen from Poona, India.

Name and notes Named after American geologist and ethnologist John Wesley Powell (1834–1902), director of the U.S. Geological Survey. First described in the Peacock lode, Seven Devils District, Idaho.

Left and above: Specimen from Ajanta, India.

Scheelite

132

Formula $CaWO_4$
System Tetragonal
Habit Typically forms bipyramidal crystals; white, yellow, pale green, orange, orange-red, brown or colorless, with an adamantine luster; common in massive and granular forms.

Environment Typical of metamorphic rock, in particular amphibolites and skarn; also present in hydrothermal veins and granitic pegmatites.
Name and notes Named after Swedish chemist Karl Scheele (1742–86), who discovered the presence of tungstate oxide in this mineral. Scheelite forms an isomorphic series with powellite; the two have the same structure, but here tungsten predominates over molybdenum.

Crystal of scheelite from the People's Republic of China.

Stolzite

133

2.5–3

7.90
8.20

Formula $PbWO_4$
System Tetragonal
Habit Forms bipyramidal and tabular crystals; brown, reddish brown, red, yellow or green, with an adamantine luster.
Environment Mineral of hydrothermal formations that forms in near-surface deposits containing minerals of lead and wolframite.
Name and notes Named after Joseph Stolz (1803–96), the collector who discovered it near Cínovec, Bohemia, Czech Republic.

Tabular crystals of stolzite from Fomarco, Piedmont, Italy.

Wulfenite

2.5–3

6.50
7.00

Formula $PbMoO_4$
System Tetragonal
Habit Forms tabular, bipyramidal and pyramidal crystals; yellow, orange, red, gray or brown, with a vitreous or adamantine luster.
Environment Mineral of hydrothermal veins that forms in near-surface deposits containing minerals of lead and molybdenum; more rarely occurs in certain granitic pegmatites.
Name and notes Named after Austrian mineralogist Franz Wülfen (1728–1805). First described at Bleiberg, in Carinthia, Austria. Wulfenite may be rich in tungsten; that is, it may show a shift in composition toward stolzite.

Top: Exceptional specimen of wulfenite from Mexico.

Above: Specimen from Arizona.

PHOSPHATES, ARSENATES, VANADATES

Class 8

APATITE GROUP

Around 30 phosphates, arsenates and vanadates belong to this group. They have a hexagonal symmetry and a general formula of the type $A_5[XO_4]_3(B)$ (in which A = Ba, Ca, Na, Pb, Sr; X = As, P, V; B = F, OH). The species illustrated here frequently present well-formed crystals in a broad range of colors, making them highly valued by museums and collectors throughout the world.

Clear yellow crystal of apatite from Mexico.

Fluorapatite

135

5

3.10
3.25

Crystal of fluorapatite from Elba, Italy.

Formula $Ca_5(PO_4)_3F$
System Hexagonal
Habit Forms hexagonal and prismatic crystals with flat or faceted bipyramidal terminations; green, violet, blue, pink, yellow, brown, white and colorless, with a vitreous luster; less common are disk-shaped and tabular crystals, and those with complex forms.
Environment Accessory of many intrusive rocks, such as syenites, granites and carbonatites; also present in metamorphic rocks, such as marble and skarn; also somewhat common in hydrothermal veins; the principal component of some sedimentary rocks known as phosphorites.
Name and notes Combines *fluor*, referring to its fluoride content, and the Greek *apatan*, "to deceive," since it was often confused for other minerals of similar appearance.

Hydroxylapatite

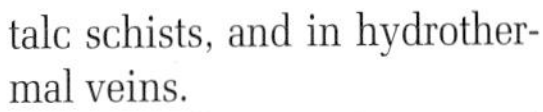

Formula $Ca_5(PO_4)_3(OH)$
System Hexagonal
Habit Usually forms prismatic crystals, hexagonal or tabular with complex forms; white, yellow, green, brown or colorless, with a vitreous luster.
Environment Far less common than fluorapatite, occurs in metamorphic rocks, notably talc schists, and in hydrothermal veins.
Name and notes Composed of the Greek words *hydor*, "water," and *apatan*, "to deceive."

Left and below: Crystals of hydroxylapatite from gneiss quarries in the Formazza Valley, Piedmont, Italy.

Mimetite

137

3.5-4 7.25

Formula $Pb_5(AsO_4)_3Cl$
System Hexagonal
Habit Forms prismatic, acicular and tabular crystals; yellow, brownish yellow, yellow-orange, white or colorless, with adamantine or resinous luster; also present in mammillary, globular and stalactitic forms.
Environment Characteristic of surficial deposits rich in minerals of arsenic and lead.
Name and notes From the Greek *mimetes*, "imitator," referring to its resemblance to pyromorphite.

Mimetite crystals from England.

Pyromorphite

138

Specimen of pyromorphite from the People's Republic of China.

Formula $Pb_5(PO_4)_3Cl$
System Hexagonal
Habit Forms typical prismatic, acicular or, more rarely, tabular crystals; green, yellow, orange-yellow, reddish orange, brownish yellow, gray or colorless, with an adamantine or resinous luster; also found in mammillary, globular, stalactitic or granular forms.
Environment Secondary mineral, characteristic of surficial deposits rich in minerals of lead; occurs rarely as product of volcanic sublimation.
Name and notes From the Greek words *pyr*, "fire," and *morphe*, "form," referring to the shapes assumed by this mineral when cooling after being melted.

Vanadinite

2.5–3 6.88

Formula $Pb_5(VO_4)_3Cl$
System Hexagonal
Habit Forms prismatic and hexagonal crystals; orange-red, bright red, yellowish brown or yellow, with adamantine or resinous luster; also common in acicular or globular aggregates, sometimes cavernous in the center.
Environment Secondary mineral, present in surficial deposits rich in minerals of lead.
Name and notes Named for its vanadium content. First described near Zimapan, Mexico.

Left and below: Specimens of vanadinite from Morocco.

AUTUNITE GROUP

This group includes roughly 15 hydrous phosphates, arsenates and vanadates of uranium, barium, calcium, copper and other elements, with tetragonal symmetry and very similar morphologies. Typical are micaceous crystals in lively colors, most of all in yellow and green hues.

Autunite crystals from Autun, France.

Autunite

140

2-2.5

3.05
3.20

Formula
$Ca(UO_2)_2(PO_4)_2 \cdot 10\text{-}12H_2O$
System Tetragonal
Habit Forms tabular crystals with rectangular or octagonal contours; lemon-yellow, greenish yellow and pale green, with a vitreous luster.
Environment Secondary mineral that forms by the alteration of minerals of uranium; also present in hydrothermal veins and in granitic pegmatites.
Name and notes Named for Autun, in the Loire, France, where it was discovered. Autunite emits bright fluorescence under ultraviolet light.

Tabular crystals of autunite from Brazil.

Torbernite

Formula
$Cu(UO_2)_2(PO_4)_2 \cdot 8\text{-}12H_2O$
System Tetragonal
Habit Forms tabular crystals with rectangular or octagonal contours, also pyramidal; green or yellowish green, with a vitreous luster.
Environment Secondary mineral that forms by the alteration of minerals of uranium; also present in hydrothermal veins and granitic pegmatites.
Name and notes Named after Swedish chemist Torbern Bergmann (1735–84), Professor of Chemistry and Physics at the University of Uppsala. First described near Jáchymov, Bohemia, Czech Republic.

Below: Specimen of torbernite from near Cuneo, Italy.

Bottom: Specimen from Musonoi, Shaba, Democratic Republic of Congo.

VIVIANITE GROUP

About ten arsenates and hydrous phosphates belong to this group, with monoclinic symmetry and a general formula of the type $A_3[XO_4]\cdot 8H_2O$, in which A can represent cobalt, iron, magnesium, manganese, nickel and zinc and X arsenic and phosphorus. These are secondary minerals, usually with vivid colors, that originate through the alteration and transformation of primary sulfarsenides or phosphates.

Exceptional emerald-green prismatic crystal of vivianite, partially transparent, from Morococala, Bolivia.

Erythrite

142

1.5–2.5 | 3.06

Formula $Co_3[AsO_4]_2\cdot 8H_2O$
System Monoclinic
Habit Rare in well-formed, flat, elongate crystals, prismatic and striated; purple and pink, with an almost adamantine luster; common in radiating and fibrous aggregates.
Environment Secondary mineral that forms in surficial deposits containing minerals of arsenic, cobalt and nickel.
Name and notes From the Greek *erythros*, "red," alluding to its typical color. First described near Schneeberg, Germany.

Specimen of flat prismatic crystals of erythrite from Bou Azzer, Morocco.

1.5–2

2.68

Vivianite

Formula $Fe^{2+}_3(PO_4)_2 \cdot 8H_2O$
System Monoclinic
Habit Forms prismatic crystals, somewhat flat; blackish blue, dark blue, pale green or colorless, with a vitreous luster; also common in massive and earthy forms.
Environment Secondary mineral of metalliferous deposits and granitic pegmatites; also found as a product of replacement of organic material in fossilized bones.
Name and notes Named after British mineralogist and geologist Jeffrey G. Vivian, who discovered this mineral. First found near St. Agnes in Cornwall, England.

Above and below: Vivianite crystals from Morococala, Bolivia.

Adamite

144

3.5

4.32
4.48

Formula $Zn_2AsO_4(OH)$
System Orthorhombic
Habit Forms aggregates of radiating crystals or rosettes, not common in isolated, elongate crystals; yellow, brownish yellow, reddish green, pink, violet, blue or white, with a vitreous luster.
Environment Mineral of hydrothermal veins, of secondary origin; occurs in deposits containing other secondary minerals of arsenic and zinc.
Name and notes Named after French mineralogist Gilbert Adam (1795–1881), who discovered this mineral. First identified near Chanarchillo, Chile.

Above and right: Specimens of adamite from Mapimi, Mexico.

Amblygonite – Montebrasite

145

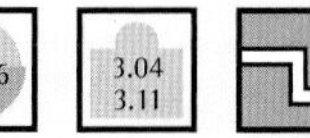

Formulas $(Li,Na)AlPO_4(F,OH)$ and $(Li,Na)AlPO_4(OH,F)$
System Triclinic
Habit Crystals, somewhat rare, are prismatic or tabular, commonly twinned; pale brown, yellow, gray, pink, white or colorless, with a vitreous luster; more commonly forms massive and granular aggregates.
Environment Typical of granitic pegmatites rich in lithium minerals and in hydrothermal veins associated with minerals of tin.
Name and notes Amblygonite is from the Greek *ambly* "blunt," and *gonios*, "angle," referring to its nearly 90° planes of cleavage. Montebrasite is named after Montebras, France, where it was discovered. Both form isomorphic mixtures: the fluorine content is dominant in amblygonite.

Amblygonite from Brazil.

Beryllonite

146

5.5-6 | 2.77 2.84

Formula $NaBe(PO_4)$
System Monoclinic
Habit Forms tabular or, rarely, prysmatic crystals, commonly twinned to simulate a hexagonal prism; white, pale yellow or colorless, with a vitreous luster; more common in fibrous and massive aggregates.
Environment Rare secondary mineral, occurs in granitic pegmatites.
Name and notes Named for its beryllium content. First found at Stoneham, Maine.

Crystal of beryllonite from Pakistan.

Brazilianite

Formula $NaAl_3(PO_4)_2(OH)_4$
System Monoclinic
Habit Forms typical wedge-shaped crystals; yellow, green-yellow or colorless, with a vitreous luster.
Environment Occurs in granitic pegmatites in zones rich in phosphates.
Name and notes Named after Brazil, where it was first found in Conselheiro Pena in the state of Minas Gerais. Highly valued by collectors for its color and the shape of its crystals.

Left: Specimen of brazilianite from Brazil that is exceptional in the color and size of its crystals.
Below: Specimen from Brazil.

Cafarsite

5.5-6 | 3.90

Formula $(Ca_8(Ti,Fe^{2+},FE^{3+},Mn)_{6\_7}(As^{3+}O_3)_{12}\cdot 4H_2O$
System Cubic
Habit Forms octahedral, cubic and more complex crystals; dark brown or reddish brown, with a vitreous luster in unaltered specimens.
Environment Rare mineral, occurs in certain alpine-type hydrothermal veins.
Name and notes The name reflects its chemical composition: calcium, iron (Latin *ferrum*) and arsenic. Discovered on Mount Leone, Switzerland.

Specimen of cafarsite from the Italian slope of Mount Cervandone.

Descloizite

3-3.5 | 6.20

Formula $Pb(Zn,Cu)VO_4OH$
System Orthorhombic
Habit Forms pyramidal crystals, prismatic or tabular; dark brown, reddish brown or orange-red, with a resinous luster; also common in stalactitic, globular and fibrous aggregates.
Environment Mineral of secondary origin that occurs in surficial deposits rich in minerals of vanadium, formed by the circulation of hydrothermal solutions.
Name and notes Named after Alfred Des Cloizeaux (1817–97), Professor of Mineralogy at the National Museum of Natural History, Paris who was first to describe this mineral.

Complex crystals of descloizite from the Berg Aukas mine, Namibia.

Herderite – Hydroxyl-herderite

Herderite from Pakistan.

Formulas $CaBe(PO_4)(F,OH)$ and $CaBe(PO_4)(OH,F)$
System Monoclinic
Habit Both form short prismatic crystals with many faces, commonly twinned or tabular; pale yellow, greenish yellow, violet, pink, gray, brown or colorless, with a vitreous luster; also common are spheroidal and fibrous-radiating aggregates.
Environment These rare phosphates occur in the cavities of granitic pegmatites.
Name and notes Named after Sigmund von Herder (1776–1838), a mine inspector in Freiberg, Germany. First described at Ehrenfriedersdorf, Saxony, Germany. The two minerals form isomorphic mixtures: the amount of fluorine (F) is dominant in herderite.

Hydroxyl-herderite from Brazil.

Lazulite – Scorzalite

151

5.5–6

3.12
3.24

Formulas $MgAl_2(PO_4)_2(OH)_2$ and $Fe^{2+}Al_2(PO_4)_2(OH)_2$
System Monoclinic
Habit Both form prismatic crystals, pseudobipyramidal; sky-blue, yellowish green or bluish green, with a vitreous luster; also common in massive and granular aggregates.
Environment Associated with quartz veins, occurs in metamorphic rocks and in intrusions in contact with granitic pegmatites.
Name and notes Lazulite is from the Latin *lazulum* or Persian *lazhuward*, "blue." Scorzalite is named for Brazilian mineralogist Evaristo Scorrzu, who discovered it. Both form isomorphic mixtures; the contents of magnesium is dominant in lazulite.

Top: Scorzalite from California.

Above: Lazulite from the Yukon Territory, Canada.

Legrandite

152

Formula $Zn_2(AsO_4)(OH)\cdot H_2O$
System Monoclinic
Habit Forms groups of elongated, radiating prismatic crystals, striated; canary-yellow, orange or colorless, with a vitreous luster.
Environment This somewhat rare mineral occurs in surficial zones of deposits rich in arsenic and zinc; also present in granitic pegmatites.
Name and notes Named after a Belgian mining engineer named Legrand, who collected the first specimen.

Specimen of legrandite from Mexico.

Liroconite

153

2-2.5 2.94 3.01

Formula
$CuAl(AsO_4)(OH)_4\cdot 4H_2O$
System Monoclinic
Habit Forms pseudo-octahedral lenticular crystals; sky-blue, green-blue or emerald-green, with a vitreous or resinous luster.
Environment Somewhat rare secondary mineral that occurs in surficial deposits rich in copper.
Name and notes From the Greek *leiros*, "pale," and *konis*, "powder," referring to the pale blue color of the mineral when ground.
Blue lirconite from Cornwall, England.

Lithiophilite – Triphylite

154

4-5 3.44 3.50

Formulas $LiMnPO_4$ and $LiFePO_4$
System Orthorhombic
Habit Both form compact granular masses, and more rarely prismatic crystals; grayish blue, greenish gray, brown and black, with a vitreous luster.
Environment Typical primary phosphates, develops in granitic pegmatites.
Name and notes Lithiophilite is named for the Latin *lithium* and the Greek *philos*, "friend," referring to its high lithium content. Triphylite is from the Greek *tria*, "three," and *phylon*, "family," since it contains iron, lithium and magnesium. The two form an isomorphic series in which lithiophilite is the manganese-dominant member.

Specimen from Portugal.

Ludlamite

155

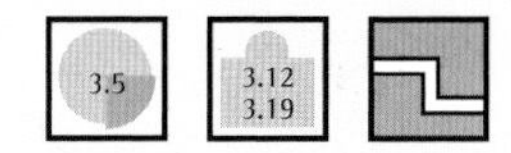

Formula $Fe^{2+}{}_3(PO_4)_2 \cdot 4H_2O$
System Monoclinic
Habit Forms typical tabular crystals joined in parallel aggregates; apple-green, with a vitreous luster.
Environment Occurs in hydrothermal veins, also found in metamorphic skarn.
Name and notes Named after British mineral collector Henry Ludlam (1824–80). First described near Wheal Jane, Cornwall, England.

Specimen of ludlamite from Mexico.

Monazite-(Ce)

5–5.5

4.89
5.43

Formula $CePO_4$
System Monoclinic
Habit Forms prismatic or tabular crystals; reddish brown, brown, pale yellow, pink, gray, yellowish brown and green, with a vitreous or resinous luster; also common in massive and granular forms.
Environment Accessory of igneous rocks, such as granites, syenites, pegmatites and carbonatites; present in metamorphic rocks and in quartz veins in hydrothermal veins.
Name and notes From the Greek *monazein*, "solitary," alluding to its rarity and to its occurrence as isolated crystals. Discovered in Zlatoust, in the Ilmen Mountains, Russia.

Above: Monazite-(Ce) crystal from Norway.
Below: Crystals from Mount Cervandone, Piedmont, Italy.

Pharmacosiderite

157

2.5

2.80

Formula
$KFe^{3+}_4(AsO_4)_3(OH)_4 \cdot 6\text{-}7H_2O$
System Cubic
Habit Common in cubic crystals; green, yellow, dark brown or reddish brown, with an adamantine luster.
Environment Typical secondary mineral produced by the alteration of sulfides containing arsenic.
Name and notes From the Greek words *pharmakon*, "drug," and *sideros*, "iron," for its use in antiquity as a medicine. First found in Cornwall, England.
Crystals from the Czech Republic.

Phosphophyllite

3-3.5

3.08
3.13

Formula
$Zn_2(Fe^{2+},Mn^{2+})(PO_4)_2 \cdot 4H_2O$
System Monoclinic
Habit Very rare, can form beautiful, complex prismatic crystals; blue-green, sea-green or colorless, with a vitreous luster.
Environment Secondary mineral of granitic pegmatites rich in phosphates; also found, but rarely, in hydrothermal veins, in deposits rich in minerals of iron and zinc.
Name and notes Named for its phosphorus content and the Greek *phyllon*, "leaf," alluding to its perfect cleavage. Valued by collectors for its color and the shape of its crystals. First found near Hagendorf, Bavaria, Germany.

Crystal from Hagendorf, Germany.

Wardite

159

Formula
$NaAl_3(PO_4)_2(OH)_4 \cdot 2H_2O$
System Tetragonal
Habit Forms bipyramidal and pseudo-octahedral crystals; brown, yellow-brown, yellow-green, pale green or colorless, with a vitreous luster.
Environment Mineral of hydrothermal veins, commonly associated with other rare phosphates; more rarely occurs in granitic pegmatites.
Name and notes Named after American collector and mineral merchant Henry Ward (1834–1906). First found near Clay Canyon, Utah.

Pseudo-octahedral crystals of wardite from the Yukon Territory, Canada.

Wavellite

160

3.5-4 2.36

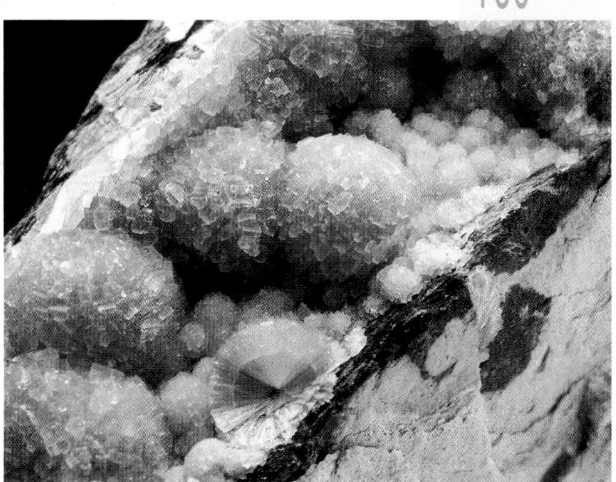

Formula
$Al_3(PO_4)_2(OH,F)_3 \cdot 5H_2O$
System Orthorhombic
Habit Forms globular aggregates of radiating crystals; green, yellowish green, greenish white or turquoise, with a vitreous or resinous luster; less common are isolated crystals, prismatic or elongate.
Environment Secondary mineral present in deposits rich in phosphates associated with metamorphic rocks.
Name and notes Named after British physician William Wavell (d. 1829), who discovered the mineral in Barnstable, Devonshire, England.

Globular aggregates of wavellite crystals from Arkansas.

Xenotime-(Y)

4-5

4.40
5.10

Formula YPO_4
System Tetragonal
Habit Forms typical prismatic and bipyramidal crystals; yellow, yellowish brown, reddish brown, red or pale green, with a vitreous or resinous luster; also common in crystals joined in radiating aggregates.
Environment Accessory mineral of granitic pegmatites; also present in quartz veins of hydrothermal formations.
Name and notes From the Greek *xenos*, "stranger," as it is rarely found and generally confused with other minerals.

Above and right: Crystals of xenotime-(Y) from Alpine fissures in the Ossola Valley, Piedmont, Italy.

SILICATES

Class 9

GADOLINITE GROUP

This group contains a series of rare silicates (fewer than ten are known) with a monoclinic symmetry and somewhat complex formulas containing such elements as boron, beryllium, calcium, yttrium, iron and the rare earths.

Gadolinite has a certain historical interest since it was among the minerals from which chemists first isolated elements belonging to the rare earths at the end of the 19th century.

Gadolinite-(Y) from Norway.

Datolite

162

5–5.5

2.96
3.00

Formula $CaBSiO_4(OH)$
System Monoclinic
Habit Forms prismatic crystals, short and many-faceted; yellow, green, pink, white or colorless, with a vitreous luster; also present in globular, granular and massive forms.
Environment Mineral of hydrothermal deposits, occurs in cavities of effusive igneous rocks, in granitic pegmatites and in metamorphic rocks like skarn and serpentinites.
Name and notes From the Greek *datysthai*, "to divide," referring to the massive aggregates, which are granular and crumble.

Specimen of datolite from Ciano d'Enza, Emilia Romagna, Italy.

Gadolinite-(Y)

Formula
$(Y,REE)_2Fe^{2+}Be_2Si_2O_{10}$
System Monoclinic
Habit Generally forms short prismatic crystals; black, greenish black, reddish brown and, more rarely, green, with a vitreous luster; more common in compact masses, glassy and fractured.
Environment Occurs in granites and alkaline granitic pegmatites; also present in alpine-type hydrothermal veins with quartz.
Name and notes Named after Finnish chemist Johan Gadolin (1760–1852).

Crystals of gadolinite-(Y) in an unusual green color and highly transparent, from a pegmatite vein in the Vigezzo Valley, Piedmont, Italy.

Hingganite-(Y)

164

5-5.5 | 4.42 4.57

Formula $(Y,Yb)_2Be_2Si_2O_8(OH)_2$
System Monoclinic
Habit Forms prismatic crystals very similar to those of gadolinite; pale green, yellowish green, white or colorless, with a vitreous luster.
Environment Occurs in cavities of alkaline granitic pegmatites; also present in alpine-type hydrothermal veins with quartz.
Name and notes From *Hinggan*, a mountain chain in the province of Heilongjiang, People's Republic of China, where the mineral was found.

Prismatic crystal of hingganite-(Y) from the granophyre quarries of Cuasso al Monte, Lombardy, Italy.

GARNET GROUP

This group comprises about 15 silicates with cubic symmetry containing calcium, iron, magnesium or manganese with aluminum, chromium, titanium and vanadium. These minerals are of great interest to petrology for what they reveal of the mechanisms behind the formation of many metamorphic and igneous rocks. The garnet group is divided into two main series, known as the pyralspites (pyrope-almandine-spessartine) and ugrandites (uvarovite-grossular-andradite).

Specimen of grossular (variety "hessonite" or cinnamon stone) from the Susa Valley, Piedmont, Italy.

Almandine

165

Formula $Fe_3Al_2(SiO_4)_3$
System Cubic
Habit Well-developed crystals, rhombododecahedral, trapezohedral or more complex combinations; deep red, violet-red or blackish red, with a vitreous luster.
Environment Widespread in metamorphic rocks, in particular mica schists, gneiss and skarn; also present in igneous rocks, such as granites and eclogites.
Name and notes From *alabandina,* the Latin name of Alabanda, an ancient city of Asia Minor that was the source of cabochon-cut agates, called *alabandic carbuncle.* The transparent varieties of almandine are cut to produce gemstones.

Specimen from the Passiria Valley, Alto Adige, Italy.

Andradite

6.5–7 | 3.80 3.90

Formula $Ca_3Fe^{3+}{}_2(SiO_4)_3$
System Cubic
Habit Forms crystals that are trapezohedral, rhombododecahedral, or more complex combinations of forms; black, yellowish brown, reddish brown, green, greenish yellow or yellow, with a resinous to adamantine luster.
Environment Metamorphic rocks, such as skarn, schists and serpentinites; also found in alkaline igneous rocks.
Name and notes Named after Brazilian statesman and geologist José Bonifacio d'Andrada (1763–1838), who was the first to study it.
Above: Specimen from Greece.
Right: Specimen from the Malenco Valley, Lombardy,

Grossular

167

Formula $Ca_3Al_2(SiO_4)_3$
System Cubic
Habit Crystals are commonly rhombododecahedral, trapezohedral or more complex combinations of forms; yellow-green, green, yellow, pink, red, orange, brownish red, yellowish brown, white or colorless, with a vitreous or resinous luster.
Environment Mineral characteristic of metamorphic settings, associated with skarn, rodingites and serpentinites.
Name and notes From New Latin *grossularia*, "gooseberry," for the resemblance of its green crystals to that fruit. The transparent and colored varieties are cut to produce gemstones.

Grossular crystals from the Valle d'Aosta, Italy.

Pyrope

7-7.5 3.58

Formula $Mg_3Al_2(SiO_4)_3$
System Cubic
Habit Forms trapezohedral or rhombododecahedral crystals; reddish violet, pinkish red, orange-red, pink to almost colorless, with a vitreous luster; also common in granular and massive forms.
Environment Occurs in ultramafic igneous rocks, such as peridotites, kimberlites and eclogites; also present in metamorphic rocks, in particular quartz schists, amphibolites and serpentinites.
Name and notes From the Greek *pyropos*, "fire-eyed," in reference to its characteristic red color.

Pyrope crystal from Piedmont, Italy.

Spessartine

169

Specimen of spessartine from the People's Republic of China.

Formula $Mn_3Al_2(SiO_4)_3$
System Cubic
Habit Forms rhombododecahedral and trapezohedral crystals, and combinations of more complex forms, often well developed; red, orange-red, orange, yellow, yellowish brown to reddish brown, with a vitreous luster.
Environment Common in granitic pegmatites, granites and rhyolites; also found in skarn and in deposits of manganese associated with metamorphic rocks.
Name and notes Named after Spessart, Germany, where it was first found. Spessartine forms an isomorphic series with almandine and pyrope in which, maintaining the same crystalline structure, the contents of manganese (Mn), iron (Fe) and magnesium (Mg) vary.

Uvarovite

170

6.5-7 | 3.77 3.81

Formula $Ca_3Cr_2(SiO_4)_3$
System Cubic
Habit Found in rhombododecahedral and trapezohedral crystals, and in combinations of more complex forms; emerald-green or dark green, with a vitreous luster.
Environment The rarest of the garnets illustrated here, it is found in metamorphic rocks such as serpentinites and metamorphic skarn.
Name and notes Named after Count Sergei S. Uvarov (1786–1855), Russian statesman and president of the St. Petersburg Academy of Sciences. First described in Saransky, Russia. Uvarovite forms an isomorphic series with andradite and grossular in which, maintaining the same crystalline structure, the contents of chromium (Cr), iron (Fe) and aluminum (Al) vary.

Uvarovite specimen from Norway.

OLIVINE GROUP

This group includes four silicates with an orthorhombic symmetry and a general formula of the type A_2SiO_4 (in which A = Fe, Mg, Mn, Ni). Fayalite and forsterite frequently form a solid solution (isomorphic series) and for that reason are generically defined by the term olivine. Olivine is believed to be among the main constituents of the earth's mantle.

Nodule of fayalite.

Fayalite

171

6.5–7

4.39

Formula Fe_2SiO_4
System Orthorhombic
Habit Forms tabular crystals with cuneiform terminations; greenish yellow, green, brown-yellow or brown, with a vitreous luster; also common in massive and granular forms.
Environment Occurs in alkaline syenitic or granitic igneous rocks; also present in metamorphic rocks with iron-rich composition.
Name and notes Named for the island of Fayal in the Azores, Portugal, where it was discovered in 1840.

Nodule of fayalite contained in granophyre from Cuasso al Monte, Lombardy, Italy.

Forsterite

7

3.27

Formula Mg_2SiO_4
System Orthorhombic
Habit Forms idiomorphic crystals terminating in bipyramidal faces; green, yellow, white, gray or grayish blue, with a vitreous luster; also common in massive and granular forms.
Environment Occurs in ultramafic igneous rock and in metamorphic rocks of ultramafic composition.
Name and notes Named after German naturalist Johann R. Forster (1729–98), who sailed with Capt. James Cook and discovered forsterite on Mount Somma-Vesuvius. Fayalite and forsterite form an isomorphic series, maintaining the same crystalline structure, but vary in proportions of magnesium and iron.

Top: Specimen from Pakistan.
Above: Specimen from Sri Lanka.

Andalusite

173

6.5
7.5

3.13
3.16

Formula $Al^{[6]}Al^{[5]}OSiO_5$
System Orthorhombic
Habit Forms elongate prismatic crystals with square cross-section; pink, pinkish brown, violet, yellow, green or gray, with a vitreous luster; also common in massive and fibrous forms.
Environment Characteristic of metamorphic schistose rocks; occurs more rarely in granites and pegmatites.
Name and notes Named after the Spanish region of Andalusia, where it was first found.

Prismatic crystals of andalusite from the Bregaglia Valley, Italy.

Braunite

174

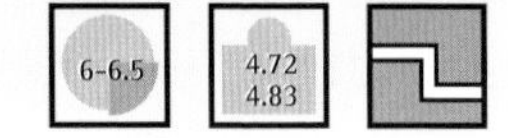

Formula $Mn^{2+}Mn^{3+}{}_6SiO_{12}$
System Tetragonal
Habit Infrequent in bipyramidal crystals; black, steel-gray to brownish black, with a metallic luster; more common in massive and granular aggregates.
Environment Occurs in metamorphic skarn, commonly associated with other silicates and oxides of manganese.
Name and notes Named after German mineral collector Kamerath Braun (1790–1872).

Crystal of braunite from Praborna, Valle d'Aosta, Italy.

175

7.5

2.99
3.10

Euclase

Formula $BeAlSiO_4(OH)$
System Monoclinic
Habit Forms prismatic crystals, commonly well-terminated; colorless, white, yellowish green, green, blue and greenish blue, with a vitreous luster.
Environment Rare mineral, found in granitic pegmatites and alpine-type hydrothermal quartz veins.
Name and notes From the Greek words *eu*, "good," and *klas*, "to break," alluding to its perfect cleavage.

Specimen of euclase from Brazil.

176

5.5

3.53
3.65

Kyanite

Formula Al_2SiO_5
System Triclinic
Habit Forms laminar, tabular and elongate crystals; blue, white, green, gray, yellow, pink or colorless, with a vitreous or pearly luster.
Environment Found in metamorphic rocks like gneiss and schist, commonly associated with quartz veins.
Name and notes From the Greek *kyanos*, "blue," alluding to its characteristic color.

Clear crystal of kyanite from Mount Forno, Switzerland.

Phenakite

Formula Be_2SiO_4
System Trigonal
Habit Forms prismatic and rhombohedral crystals and more complex forms, more rarely forms prismatic acicular crystals; colorless, white, yellow, pink or brown, with a vitreous luster.
Environment Occurs in granitic pegmatites and alpine-type hydrothermal veins.
Name and notes From the Greek *phenax*, "deceiver," as it was mistaken for other minerals of similar appearance.

Crystals from Brazil.

Sapphirine

178

7.5

3.40
3.58

Formula $(Mg,Al)_8(Al,Si)_6O_{20}$
System Triclinic or monoclinic
Habit Forms rare idiomorphic crystals; blue, sky-blue, green-blue, white, gray and yellow, with a vitreous luster; more commonly found in granular and massive aggregates.
Environment Associated with metamorphic rocks rich in aluminum and magnesium; less common in intrusive mafic igneous rocks.
Name and notes From the Greek *sapphire*, in reference to the characteristic deep blue color of this mineral.

Centimetric prismatic crystals of sapphirine from the area of Fort Dauphin, in southern Madagascar.

Sillimanite

179

6.5–7.5 | 3.23 3.24

Formula $^{[6]}Al^{[4]}AlSiO_5$
System Orthorhombic
Habit Forms prismatic, acicular crystals with square cross-section, striated; white, gray, yellow-green, green, gray-green, blue-green, blue or colorless, with vitreous luster; also common in fibrous aggregates.
Environment Typical of metamorphic rocks such as schist, gneiss and hornfels; occurs rarely in pegmatites.
Name and notes Named after Benjamin Silliman (1779–1864), first Professor of Mineralogy at Yale University. First described near Chester, Connecticut. Sillimanite, andalusite and kyanite are polymorphs, having the same chemical composition but different crystalline structures.

Crystal of sillimanite from the Ossola Valley, Piedmont, Italy.

Staurolite

180

7–7.5 | 3.74 3.83

Formula $(Fe,Mg)_{3\text{-}4}(Al,Fe)_{18}(Si,Al)_8O_{48}H_{2\text{-}4}$
System Monoclinic
Habit Characteristic prismatic crystals, and twin crystals commonly form a right-angle cross; reddish brown, blackish brown or yellowish brown, with a resinous or vitreous luster.
Environment Typical of metamorphic rocks such as schist and gneiss.
Name and notes Name is from the Greek *stauros*, "cross," for its characteristic crosslike twinned form.

Crystal contained in schist from Mount Forno, Switzerland.

Titanite

181

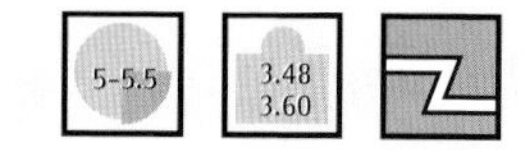

Formula $CaTiOSiO_4$
System Monoclinic
Habit Common in wedge-shaped crystals, somewhat flattened, or prismatic; brown, gray, green, yellow, red or colorless, with an adamantine to resinous luster.
Environment Accessory mineral of many intrusive rocks, pegmatites and alpine-type hydrothermal veins; also present in metamorphic rocks, such as schist, gneiss and skarn.
Name and notes Named for its titanium content.

Flat, wedge-shaped crystal of titanite associated with hydroxylapatite from an alpine-type hydrothermal vein in Ankarafa, Madagascar.

Topaz

8

3.49
3.57

Formula $Al_2SiO_4(F,OH)_2$
System Orthorhombic
Habit Forms beautiful prismatic crystals, commonly well-terminated; yellow, pink, red, orange, brown, green, sky-blue, blue, violet or colorless, with a vitreous luster.
Environment Occurs in granites and in the cavities of granitic pegmatites, in rhyolites and in high-temperature hydrothermal veins of quartz (known as *greisen*), in association with cassiterite, fluorite, quartz and mica.
Name and notes Probably from the Sanskrit *tapas*, "fire," reflecting this mineral's brilliance. The transparent colored varieties are cut to produce gemstones of great value.

Topaz crystal from Pakistan.

Uranophane

183

2.5

3.81
3.90

Uranophane crystals from Bavaria, Germany.

Formula $Ca(UO_2)_2[SiO_3(OH)]_2 \cdot 5H_2O$
System Monoclinic
Habit Forms prismatic, acicular crystals, commonly in radiating aggregates; lemon-yellow, brown-yellow, greenish yellow or orange-yellow, with a vitreous luster; also forms thin microcrystalline and fibrous incrustations.
Environment Secondary mineral that forms in uranium deposits and in granitic pegmatites from the alteration of uraninite.
Name and notes From *uranium* and the Greek *phainesthai,* "to appear," reflecting the uncertainty of its composition when the mineral was first analyzed. First found near Miedzianka in Silesia, Poland.

Zircon

184

7.5

4.60
4.70

Zircon crystals from Malawi.

Formula $ZrSiO_4$
System Tetragonal
Habit Forms typical prismatic or tabular crystals with square cross-sections, terminating in bipyramids; reddish brown, red, orange, yellow, green, gray, blue or colorless, with a vitreous or adamantine luster; also common in granular and massive forms.
Environment Accessory mineral of many metamorphic and igneous rocks; can also form large deposits of alluvial detritus.
Name and notes From the Persian *zar*, "golden," and *gun*, "colored," alluding to one of the many colors of this mineral. The transparent varieties are cut to produce gemstones.

AXINITE GROUP

This small group of silicates (only four minerals are known) with triclinic symmetry are hydrous and contain calcium, iron, magnesium or manganese. The minerals in this group form partial or complete isomorphic solutions; they have very complex crystalline forms with well-faceted crystals and characteristic sharp edges.

Specimen of ferro-axinite from the Oisans district of France.

Ferro-axinite

185

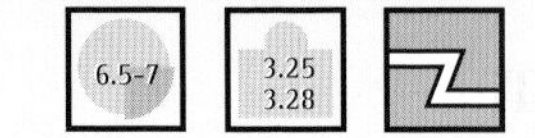

Formula $Ca_2Fe^{2+}Al_2BO(OH)(Si_2O_7)_2$
System Triclinic
Habit Forms typical flattened, wedge-shaped crystals; clove-brown, with a vitreous luster.
Environment Mineral of alpine-type hydrothermal veins; also common in metamorphic skarn and pegmatites.
Name and notes From the Latin *ferro*, "iron," because of the presence of that element, and the Greek *axine*, "ax," alluding to the typical shape of its crystals.
Crystals from the Ural Mountains, Russia.

Manganaxinite

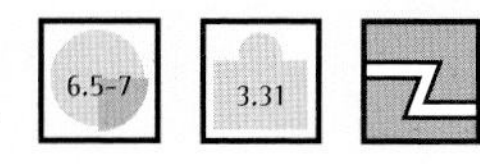

Formula
$Ca_2Mn^{2+}Al_2BO(OH)(Si_2O_7)_2$
System Triclinic
Habit Forms typical flattened, wedge-shaped crystals; honey-yellow, clove-brown or brown, with a vitreous luster.
Environment Mineral of alpine-type hydrothermal veins; also common in metamorphic skarn and pegmatites.
Name and notes From *mangan* for its quantity of manganese, and *axine*, Greek for "ax." Manganaxinite forms an isomorphic series with ferroaxinite, which has the same crystalline structure, but in manganaxinite the manganese content is dominant.

Typical flattened crystals of manganaxinite from Arizona.

Tinzenite

187

6.5
3.35
3.43

Formula
$CaMn^{2+}{}_2Al_2BO(OH)(Si_2O_7)_2$
System Triclinic
Habit Found in groups of flattened, lenticular prismatic crystals; yellow, brown or yellow-green, with a vitreous luster.
Environment Typical of manganese-rich deposits associated with metamorphic rocks.
Name and notes Named after Tinzen, Switzerland, where it was discovered. Tinzenite forms an isomorphic series with manganaxinite, maintaining the same crystalline structure, but in tinzenite the calcium content is dominant.

Tinzenite from a manganese mine in Liguria, Italy.

EPIDOTE GROUP

Around 15 silicates are included in this group, characterized by monoclinic or orthorhombic symmetry with the general formula $A_2B_3(SiO_4)_3(OH)$, in which (A = Ca, Ce, Pb, Sr, Y, and B = Al, Fe, Mg, Mn, V). Epidote is a relatively common mineral in nature since it is part of the composition of numerous igneous and metamorphic rocks. Except for those illustrated here, the members of this group are relatively rare.

Group of epidote crystals in which the prismatic habit is quite plain to see, from alpine-type hydrothermal fissures in the Himalayan chain in Pakistan.

Allanite-(Y)

188

5.5–6

3.50
4.20

Formula $Ca(Y,LaCe)(Al, Fe^{3+})_3(SiO_4)_3(OH)$
System Monoclinic
Habit Forms tabular, prismatic or acicular crystals; black or brown, with a vitreous or metallic luster; also common in granular and massive aggregates.
Environment Occurs in intrusive igneous rocks, such as granites, syenites and granitic pegmatites; less common in metamorphic rocks like gneiss and skarn.
Name and notes Named after Scottish mineralogist Thomas Allan (1777–1833), who was the first to recognize this mineral.

Allanite-(Y) crystal from a talc quarry in Trimouns, France.

Epidote

189

6.5–7

3.38
3.49

Formula
$Ca_2Al_2(Fe^{3+},Al)Si_3O_{12}(OH)$
System Monoclinic
Habit Forms elongate, tabular crystals and combinations of prismatic forms, generally terminating in sloping faces; green, yellowish green, greenish yellow or greenish black, with a vitreous luster; also common in fibrous, granular and massive aggregates.
Environment Accessory of many metamorphic rocks and intrusive igneous rocks.
Name and notes From the Greek *epidosis*, "to increase," alluding to the typical elongate shape of its crystals.

Above: Specimen of epidote from veins of rodingite in the Susa Valley, Piedmont, Italy. Left: Specimen from Prince of Wales Island, Alaska.

Piemontite

190

6-6.5 | 3.46 3.54

Formula $Ca_2(Al,Mn^{3+},Fe^{3+})_3Si_3O_{12}(OH)$
System Monoclinic
Habit Forms prismatic, laminar and acicular crystals, in some cases joined in radiating groups; reddish brown, dark red, violet-red or violet, with a vitreous luster.
Environment Occurs in deposits rich in manganese, associated with schistose, amphibole-rich metamorphic rocks; also present in hydrothermal veins associated with rhyolites, diorites and andesites.
Name and notes Named after the Piedmont region of Italy by English mineralogists in the 19th century; the mine of Praborna near San Marcel, where it was first found, later became part of the Val d'Aosta region.

Specimen of piemontite from Praborna, Italy.

Zoisite

191

6-7 | 3.15 3.36

Formula $Ca_2Al_3Si_3O_{12}(OH)$
System Orthorhombic
Habit Forms prismatic crystals, generally striated; gray, white, greenish brown, pink, blue or violet, with a vitreous luster; also common in columnar and massive aggregates.
Environment Forms in metamorphic rocks rich in silicates of calcium, in particular rodingites and eclogites.
Name and notes Named after Baron S. Zois van Edelstein (1747–1819), Austrian collector and financier of numerous mineralogical expeditions. Discovered near Saualpe, in Carinthia, Austria.

Left and right: Prismatic crystals of zoisite from the Shigar Valley, Pakistan.

Barylite

192

7

4.07

Formula $BaBe_2Si_2O_7$
System Orthorhombic
Habit Thin, tabular prismatic crystals with square cross-section; white or colorless, with a vitreous luster.
Environment Rare mineral occurring in metamorphic deposits rich in minerals of zinc and in intrusive igneous rocks, in particular granitic pegmatites.
Name and notes From the Greek *barys,* "heavy," in allusion to its high density. First found at Långban, in Värmland, Sweden.

Above and left: Exceptional tabular crystals of barylite from Malawi.

Bertrandite

Formula $Be_4Si_2O_7(OH)_2$
System Orthorhombic
Habit Forms thin tabular crystals, prismatic, commonly twinned in a V; yellow or colorless, with a vitreous luster.
Environment Occurs in alpine-type hydrothermal veins and in cavities in granitic pegmatites, generally as the product of the alteration of beryl.
Name and notes Named for the French mineralogist Emile Bertrand (1844–1909), who discovered the mineral at Petit-Port, in the Loire, France.
Crystal from Annaberg, Germany.

Chevkinite-(Ce)

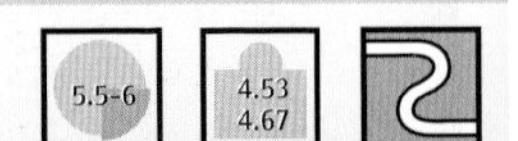

Formula $(Ce,La,Ca)_4(Fe^{2+},Mg)_2(Ti,Fe^{+3})_3Si_4O_{22}$
System Monoclinic
Habit Forms laminar and prismatic crystals; black or dark red, with a resinous luster; also common in granular and massive aggregates.
Environment Accessory of granites, syenites and alkaline pegmatites; also found in metamorphic skarn.
Name and notes Named after the Russian general Konstantin Chevkin (1802–75). First found in the Ilmen Mountains, Russia.

Specimen from Pakistan.

Hemimorphite

195

4.5–5 | 3.47

Formula $Zn_4Si_2O_7(OH)_2 \cdot H_2O$
System Orthorhombic
Habit Forms groups of tabular crystals, striated, in fan-shaped aggregates; white, blue, pale green, gray, brown or colorless, with vitreous luster; also common in fibrous, mammillary and stalactitic aggregates.
Environment Secondary mineral typical of surficial deposits rich in minerals of zinc.
Name and notes From the Greek words, *hemi*, "half," and *morphe*, "form," as the opposite ends of hemimorphite crystals have different crystal forms.

Above: Crystals from Mexico.

Below: Beautiful mammillary aggregate from the People's Republic of China.

Ilvaite

196

5.5-6

3.99
4.05

Formula
Ca $Fe^{3+}(Fe^{2+})_2O(Si_2O_7)OH$
System Monoclinic
Habit Prismatic crystals, striated; black or blackish gray, with a metallic luster; also common in columnar, radiating and massive aggregates.
Environment Typical of metamorphic skarn containing sulfides and iron oxides.
Name and notes From the Latin *Ilva,* the ancient Roman name for the island of Elba, Italy. First described at Rio Marina on Elba.

Specimen of ilvaite from Dalnegorsk, Russia.

Thortveitite

197

6-7

3.27
3.58

Formula $(Sc,Y)_2(Si_2O_7)$
System Monoclinic
Habit Forms elongate prismatic crystals; black or greenish gray, rarely whitish, with a vitreous luster.
Environment A somewhat rare mineral found in granitic pegmatites.
Name and notes Named after Norwegian engineer Olaus Thortveit, who discovered it near Iveland, Norway.

Elongate prismatic crystals found in granite at Baveno, Piedmont, Italy.

Vesuvianite

198

6-7

3.32
3.43

Formula $Ca_{19}(Al,Mg,Fe)_{13}(Si_{18}O_{68})(O,OH,F)_{10}$
System Tetragonal
Habit Common in prismatic crystals, generally striated, with complex terminations; yellow, green, brown, violet, bluish green, pink, red, black or colorless, with a vitreous luster; common also in granular, columnar and massive aggregates.
Environment Found in metamorphic skarn, in rodingites, and in serpentinites; also present in mafic, intrusive igneous rocks and in some pegmatites.
Name and notes Named after Mount Vesuvius, the volcano near Naples, Italy, where this mineral was first found.
Right: Green crystals from Asbestos, Quebec, Canada.
Below: Polychrome crystal from the Susa Valley, Piedmont, Italy.

OSUMILITE GROUP

This group includes 15 or so very rare alkaline silicates, characterized by highly complex formulas and crystal structures. Even the best-known members of this group, milarite and osumilite, are not widespread in nature and are of exclusively scientific interest.

Specimen of milarite with numerous crystals from the Giuf Valley, Switzerland.

Milarite

199

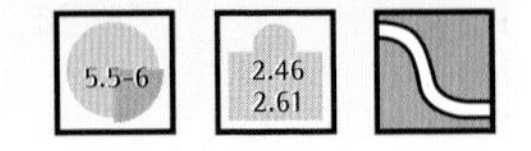

Formula
$KCa_2AlBe_2Si_{12}O_{30}{\cdot}0.5H_2O$
System Hexagonal
Habit Typical hexagonal prismatic crystals, well formed; colorless or gray, pale green, green and yellow, with a vitreous luster.
Environment Occurs in alpine-type hydrothermal veins and in granitic pegmatites.
Name and notes Named for where it was found, Val Guif (Val Milar), Tonetsch, Grischum, Switzerland.

Crystal from the Giuf Valley, Switzerland.

Osumilite

5-6

2.58
2.68

Formula $(K,Na)(Fe^{2+},Mg)_2(Al,Fe^{3+})_3(Si,Al)_{12}O_{30}$
System Hexagonal
Habit Prismatic crystals, tabular and hexagonal; black, dark blue, brown, pink or gray, with a vitreous luster.
Environment Occurs in cavities of alkaline effusive igneous rocks such as rhyolites and dacites; also present in skarn assemblages.
Name and notes Named after Osumi, Japan, where it was discovered.

Above and right: Specimens of osumilite from rhyolite, Mount Arci, Sardinia, Italy.

TOURMALINE GROUP

This group of silicates (currently numbering 13 species) have trigonal symmetry and somewhat complex formulas containing boron, calcium, sodium, iron, lithium, magnesium, manganese and aluminum. Aside from forming solid solutions, these minerals are known generically as tourmalines since some are well known to the public through their large-scale use as gemstones.

Elbaite from Pakistan.

Dravite

201

7

3.03
3.18

Formula $(Na,Ca)Mg_3Al_6(BO_3)_3[Si_6O_{18}](OH)_4$
System Trigonal
Habit Forms characteristic elongate prismatic crystals, striated; black, brown, red, yellow, blue, green, white or colorless, with a vitreous luster.
Environment Common in metamorphic rock and in intrusive igneous rocks rich in boron; more rarely present in evolved pegmatites.
Name and notes Named after the province of Drave, in Carinthia, Austria.

Crystal from Crevoldadossola, Piedmont, Italy.

Elbaite

202

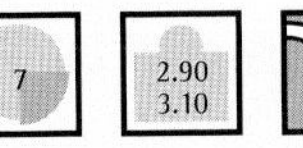

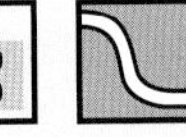

Formula $Na(Li_{1.5}Al_{1.5})Al_6(BO_3)_3[Si_6O_{18}](OH)_4$
System Trigonal
Habit Forms prismatic to acicular crystals, striated, terminating in pyramidal faces; green, blue, red, orange, yellow, colorless or polychrome, with a vitreous luster; also common in radiating, fibrous and massive groups.
Environment Characteristic of granitic pegmatites and hydrothermal veins.
Name and notes Named after the island of Elba, Italy, where it was first discovered. The transparent colored varieties are cut to produce gemstones, some of great value.

Polychrome crystal of elbaite from Pakistan.

Liddicoatite

203

Formula $Ca(Li_{1.5}Al_{1.5})Al_6(BO_3)_3[Si_6O_{18}](OH,O)_4$
System Trigonal
Habit Forms striated prismatic crystals; brown, green, pink, red or blue, with a vitreous luster.
Environment Occurs in granitic pegmatites rich in lithium.
Name and notes Named after Richard Liddicoat (1918–2002), gemologist with the Gemological Institute of America.

Crystal of liddicoatite from Anjanabonoina, Madagascar.

Schorl

204

Schorl crystals from Namibia.

Formula $NaFe^{3+}{}_3Al_6(BO_3)_3[Si_6O_{18}](OH)_4$
System Trigonal
Habit Forms prismatic, acicular or tabular crystals terminating in pyramidal faces; black or brown-black, with a vitreous luster; also common in radiating, fibrous and massive groups.
Environment Characteristic mineral of granitic pegmatites, hydrothermal veins and skarn assemblages.
Name and notes From the German mining term *Schörl* or *Schirl*, used since the 1500s for a black mineral found as waste after washing and separation from gold or cassiterite.

Uvite

205

7.5

2.97
3.14

Formula $CaMg_3Al_6(BO_3)_3[Si_6O_{18}][(OH)_3O]$
System Trigonal
Habit Forms prismatic to tabular crystals with pyramidal terminations; brown, green, red or black, with a vitreous luster.
Environment Occurs in metamorphic rocks rich in boron and calcium; more rarely occurs in granitic pegmatites.
Name and notes Named after the province of Uva in Sri Lanka, where it was first identified.

Prismatic crystals of uvite from the Brumado quarry, Brazil.

Benitoite

6–6.5

3.65

Formula $BaTiSi_3O_9$
System Hexagonal
Habit Forms flattened pyramidal crystals, tabular with triangular or hexagonal contours; sapphire-blue, white or colorless, with a vitreous luster.
Environment A rare mineral found in natrolite-rich veins in schistose serpentinite rocks.
Name and notes Named after the county of San Benito, California.

Above and right: Crystals of benitoite from San Benito, California.

Beryl

207

7.5–8

2.63
2.97

Formula $Be_3 Al_2Si_6O_{18}$
System Hexagonal
Habit Prismatic or tabular crystals with flat, bipyramidal or complex terminations; white, sky-blue, pale blue, greenish blue, bright green, greenish yellow, yellow, pink, peach-pink, reddish pink, red or colorless, with a vitreous luster; common in radiating, columnar, granular and compact aggregates.
Environment Develops in granites and granitic pegmatites, in basic metamorphic rocks, in hydrothermal veins and in rhyolites rich in aluminum.
Name and notes From the Greek *beryllos*, the ancient name for a blue-green gemstone. The transparent colored varieties of beryl are cut to produce gems, some of great value.

Left: Exceptional blue crystal from the Codera Valley, Lombardy, Italy.

Above: Clear yellow crystal of the "heliodor" variety from Tajikistan.

Cordierite

208

7-7.5 | 2.60 2.66

Formula $Mg_2Al_4Si_5O_{18}$
System Orthorhombic
Habit Rare as prismatic crystals, more common in granular and compact forms; blue, violet-blue, greenish, brownish yellow, gray or colorless, with a vitreous luster.
Environment Occurs in metamorphic schist, gneiss and granulite; also present in mafic igneous rocks and in granites rich in aluminum.
Name and notes Named for French geologist Pierre Louis Cordier (1777–1862), who first described this mineral.
Vitreous nodule of cordierite from Madagascar.

Dioptase

209

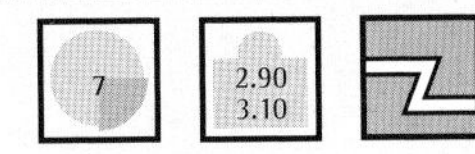

Beautiful specimen of dioptase with green rhombohedral crystals from Tsumeb, Namibia.

Formula $Cu_6Si_6O_{18}\cdot6H_2O$
System Trigonal
Habit Common in rhombohedral crystals, prismatic; emerald-green or bluish green, with a vitreous luster.
Environment Occurs in hydrothermal formations found in surficial zones of deposits rich in copper minerals.
Name and notes From the Greek words *dia,* "through," and *optasia,* "vision," referring to the fact that this mineral's cleavage planes are visible in its crystals. It was first described in 1797 at Alten-Tübe, Khirghesia, Russia, by René-Just Hauy.

Eudialyte

5-6

2.74
3.10

Formula
$Na\,Ca_6(Fe^{2+},Mn^{2+})_3Zr_3(Si,Nb)(Si_{25}O_{13})(O,OH,H_2O)_3(Cl,OH)_2$

System Trigonal

Habit Rare in rhombohedral crystals; brown, brown-yellow, yellow, pink, pinkish red or red, with a vitreous luster; more common in granular masses and in thin veins of massive aspect.

Environment Typical of intrusive alkaline rocks, such as nepheline syenites, alkaline granites and pegmatites.

Name and notes From the Greek words *eu,* "good," and *dialytos,* "dissolved," alluding to its easy solubility in acids. First found at Kangerdluarsuk, Greenland.

Above: Crystals of eudialyte from St. Hilaire, Canada.

Below: From the Kola Peninsula, Russia.

AMPHIBOLE GROUP

This vast group of silicates numbers more than 70 known species; crystals adopt a monoclinic or an orthorhombic symmetry and contain aluminum, calcium, iron, lithium, magnesium, manganese, potassium and sodium. Most of these minerals have very similar morphological characteristics and form solid, partial and complete solutions. Their study helps petrologists understand the mechanisms behind the formation of many igneous and metamorphic rocks.

Prismatic laminar crystals of actinolite from the Vizze Valley, Trentino-Alto Adige, Italy.

Actinolite

211

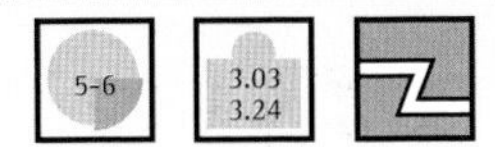

Formula
$Ca_2(Mg,Fe)_5Si_8O_{22}(OH)_2$
System Monoclinic
Habit Forms elongate and laminar crystals; green or grayish green, with a vitreous luster; also common in columnar, fibrous and radiating aggregates, both granular and massive.
Environment Accessory mineral of metamorphic rocks and mafic and ultramafic intrusive igneous rocks.
Name and notes From the Greek *aktis*, "ray," alluding to its habit common with radiating crystals.
Specimen of actinolite from Alto Adige, Italy.

Anthophyllite

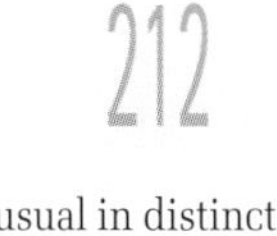

5.5-6

2.90
3.50

Formula
$(Mg,Fe^{2+})_7Si_8O_{22}(OH)_2$
System Orthorhombic

Habit Unusual in distinct laminar crystals, more common in aggregates of prismatic, lamellar or fibrous crystals; gray, brown-gray, yellowish brown, brown, green-brown or green, with a vitreous luster.
Environment Occurs in metamorphic rocks such as amphibolites, gneiss, granulites and quartz schists; also present in some intrusive mafic igneous rocks.
Name and notes From the Latin *anthophyllum*, "clove," alluding to this mineral's characteristic brown color.

Aggregate nodule of anthophyllite from the United States.

Arfvedsonite

213

5-6

3.30
3.50

Formula
$NaNa_2(Fe_4^{2+}Fe^{3+})Si_8O_{22}(OH)_2$
System Monoclinic
Habit Forms prismatic crystals, tabular and striated; dark green, with a vitreous luster; also present in radiating, prismatic and fibrous aggregates.
Environment Somewhat common accessory in granites and alkaline pegmatites.
Name and notes Named after the Swedish chemist Johan A. Arfvedson (1792–1841), who described it for the first time from the Ilímaussaq complex, Greenland.
Crystals of arfvedsonite from Malawi.

Ferrohornblende

214

5-6

3.12
3.30

Formula $Ca_2[Fe^{2+}_4(Al,Fe^{3+})](Si_7Al)O_{22}(OH)_2$
System Monoclinic
Habit Forms prismatic crystals; green to greenish brown, with a vitreous luster.
Environment Occurs in igneous rocks, like granites, granodiorites and pegmatites; accessory in metamorphic rocks, like amphibolites and schists.
Name and notes Named for its iron (*ferro*) content and for the German *horn* ("horn") and *blende* ("to deceive"), alluding to its similarity to valuable minerals found in ores.
Specimen from Cervo Valley, Piedmont, Italy.

Glaucophane

215

Formula
$Na_2(Mg_3Al_2)Si_8O_{22}(OH)_2$
System Monoclinic
Habit Forms prismatic, aggregated and columnar crystals, fibrous or granular; gray or lavender-blue, with a vitreous or pearly luster.
Environment Occurs in blue schists, metamorphic rocks indicative of a high pressure.
Name and notes From the Greek words *glaukos*, "bluish green," and *phainesthai*, "to appear," alluding to the typical color of this mineral.

Specimen of glaucophane from Piedmont, Italy.

Kaersutite

216

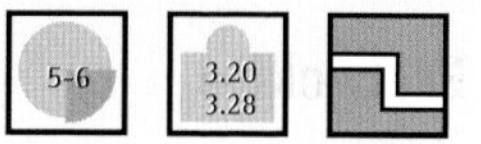

Formula
$NaCa_2(Mg_4Ti)Si_6Al_2O_{23}(OH)$
System Monoclinic
Habit Forms distinct prismatic crystals with pseudorhombic contours; dark brown or black, with a vitreous luster; also common in granular form.
Environment Common in phenocrysts in effusive igneous rocks, primarily alkali basalts; also present in intrusive igneous rocks, such as syenites, monzonites and alkaline gabbros.
Name and notes Named after Kaersut, Greenland, where the mineral was discovered.

Characteristic prismatic crystal of kaersutite from Predazzo, Trentino-Alto Adige, Italy.

Pargasite

217

5-6
3.04
3.17

Specimen with typical prismatic crystals from Sri Lanka.

Formula $NaCa_2(Mg_4Al)Si_6Al_2O_{22}(OH)_2$
System Monoclinic
Habit Typical crystals are prismatic and short; bluish green, grayish black, pale brown or emerald-green, with a vitreous luster; also common in granular form.
Environment Component of metamorphic rocks rich in calcium, such as marble and skarn; also present in schistose metamorphic rocks, in amphibolites and in some mafic igneous rocks.
Name and notes Named after Pargas, Finland, near where the mineral was first found.

Riebeckite

218

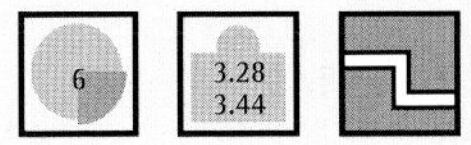

Specimen from alkaline pegmatites of Greenland.

Formula $Na_2(Fe^{2+}{}_3Fe^{3+}{}_2)Si_8O_{22}(OH)_2$
System Monoclinic
Habit Commonly forms prismatic to acicular crystals or fibrous aggregates; black or dark blue, with a vitreous luster.
Environment Occurs in intrusive igneous rocks, such as granites, syenites and alkaline pegmatites, more rarely in schist.
Name and notes Named after the German explorer and mineral collector Emil Riebeck (1853–85). First described on the island of Socotra, South Yemen.

Tremolite

219

5-6
2.99
3.03

Formula $Ca_2Mg_5Si_8O_{22}(OH)_2$
System Monoclinic
Habit Forms elongate laminar crystals; white to dark gray and pink, with a vitreous luster; also common in fibrous-radiating aggregates.
Environment Occurs in metamorphic rocks, such as marble and schist.
Name and notes Named after the Val Tremola, near St. Gotthard, Switzerland. Tremolite forms an isomorphic series with actinolite, but is dominant in magnesium.

Tremolite crystals from Russia.

MONOCLINIC PYROXENE GROUP

Roughly 15 silicates belong to this group; they have the general formula ABZ_2O_6, in which A = Na, Ca, Mn, Fe, Mg, Li; B = Al, Cr, Fe, Mg, Mn, Sc, Ti; and Z = Si, Al. Some of these minerals form solid solutions and are of great interest to the field of petrology for what they reveal about the conditions under which metamorphic and igneous rocks form.

Crystal of diopside associated with grossular and titanite from the Susa Valley, Piedmont, Italy.

Aegirine

220

6

3.50
3.60

Formula $NaFe^{3+}[Si_2O_6]$
System Monoclinic
Habit Forms distinct prismatic crystals with flat, pointed or sometimes curved terminations; dark green, greenish black, reddish brown or black, with a vitreous luster; also common in acicular and fibrous-radiating aggregates.
Environment Occurs in alkaline intrusive igneous rocks such as granites, syenites, related pegmatites and carbonatites; also present in metamorphic rocks, in particular schists, gneisses, and granulites.
Name and notes From Aegir, the Teutonic god of the sea. Its obsolete name, "acmite," is from the Greek *akme*, "point," referring to its crystals.

Prismatic crystals of aegirine from Malawi.

Augite

221

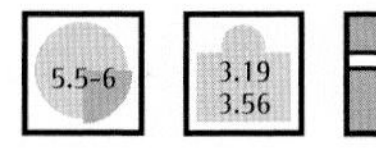

Formula $(Ca,Fe)(Mg,Fe)Si_2O_6$
System Monoclinic
Habit Forms short prismatic crystals with almost square or octagonal cross-section; black, brown, green or violet-brown, with a vitreous luster; also common in acicular and granular aggregates.
Environment Develops in mafic igneous rocks, such as basalts, andesites and gabbros; also present in some metamorphic rocks.
Name and notes From the Greek *auge*, "clear," in reference to the reflections created by the mineral's cleavage planes.

Crystals of augite from Buffaure in the Fassa Valley, Trentino-Alto Adige, Italy.

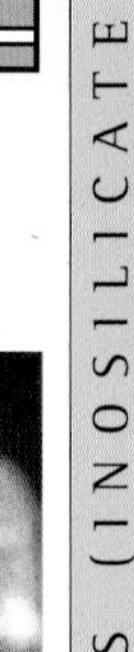

Diopside

222

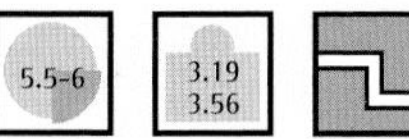

Crystal of diopside with prismatic habit from rodingite veins in the Ala Valley, Piedmont, Italy.

Formula $CaMgSi_2O_6$
System Monoclinic
Habit Forms prismatic crystals with almost square cross-section, sometimes thin; green, green-yellow, yellow or colorless, with a vitreous luster; also common in columnar and granular aggregates.
Environment Occurs in metamorphic rocks rich in calcium and magnesium; present in schist, gneiss and igneous rocks, like peridotites, kimberlites, basalts and andesites.
Name and notes From the Greek *dis*, "double," and *opsis*, "appear," as the crystals have a different habit if seen from two points of observation.

Jadeite

223

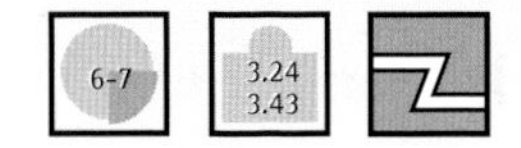

Prismatic crystals of jadeite from Arizona.

Formula $NaAlSi_2O_6$
System Monoclinic
Habit Distinct prismatic crystals are rare, the common forms being massive, fibrous and granular; apple-green, emerald-green, bluish green, greenish white or white, with a vitreous or pearly luster.
Environment Occurs in metamorphic rocks, in particular schists and, more rarely, eclogites.
Name and notes From the Spanish *piedra de ijada,* "loin stone," alluding to its ancient use as a cure for kidney ailments. Emerald-green varieties are used to produce ornamental objects.

Hedenbergite

224

Formula $Ca(Fe,Mg)Si_2,O_6$
System Monoclinic
Habit Forms prismatic crystals; dark green, brown-green or black, with a vitreous luster; also common in columnar, fibrous-radiating, granular and massive aggregates.
Environment Occurs in metamorphic rocks rich in minerals of iron; also present in some alkaline intrusive igneous rocks.
Name and notes Named after the early 19th-century Swedish chemist Ludwig Hedenberg, a student of Jöns Jacob Berzelius (1779–1848), who described the mineral. Hedenbergite forms an isomorphic series with augite and diopside, sharing the same crystal structure but with iron dominant.

Prismatic crystals from Sérifos, Greece.

Johannsenite

225

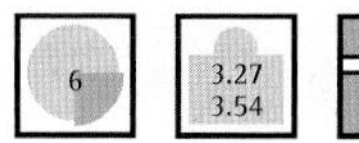

Formula $CaMnSi_2O_6$
System Monoclinic
Habit Forms prismatic crystals; brown, brown-green, gray, green or bluish green, with a vitreous luster; also common in columnar, fibrous-radiating, granular and massive aggregates.
Environment Found in deposits rich in minerals of manganese, associated with metamorphic rocks.
Name and notes Named after Albert Johannsen (1871–1962), Professor of Petrology at the University of Chicago. First described at Bohemia, Oregon.

Group of fibrous crystals from Monte Civillina in the Veneto, Italy.

Omphacite

226

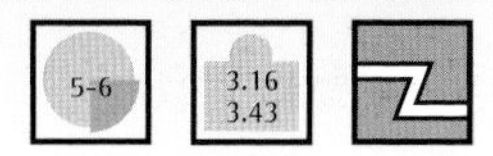

Formula $(Ca,Na)(Mg,Al)Si_2O_6$
System Monoclinic
Habit Prismatic crystals are rare, more common are granular aggregates; green, with a vitreous luster.
Environment Component of intrusive rocks like kimberlites; also present in schistose metamorphic rocks and eclogites.
Name and notes From the Greek *omphaos,* "green berry," referring to its color. First found near Hof, in Bavaria, Germany.

Crystals of omphacite from Piedmont, Italy.

Spodumene

227

6.5–7

3.03
3.23

Formula $LiAlSi_2O_6$
System Monoclinic
Habit Forms flat and striated prismatic crystals; colorless, green, greenish yellow, whitish gray, yellow, pink or violet, with a vitreous luster; also common in massive form.
Environment Found in granitic pegmatites rich in lithium.
Name and notes From the Greek *spodoumenos*, "reduce to ashes," since when heated to a high temperature this mineral becomes a grayish white mass.

Crystals of spodumene from Pakistan.

Specimen of spodumene from Brazil.

Babingtonite

228

5.5-6 | 3.34 3.37

Formula $Ca_2(Fe^{2+},Mn)Fe^{3+}Si_5O_{14}(OH)$
System Triclinic
Habit Forms somewhat flat prismatic or tabular crystals, in some cases striated; black, blackish brown or greenish black, with a vitreous luster.
Environment Found in granitic pegmatites, in the cavities of basaltic rocks and in skarn assemblages.
Name and notes Named after Irish mineralogist William Babington (1757–1833). First described in Arendal, Norway.

Group of prismatic crystals of babingtonite from cavities in basalts near Malad, India.

Bavenite

229

5.5-6 | 2.71 2.74

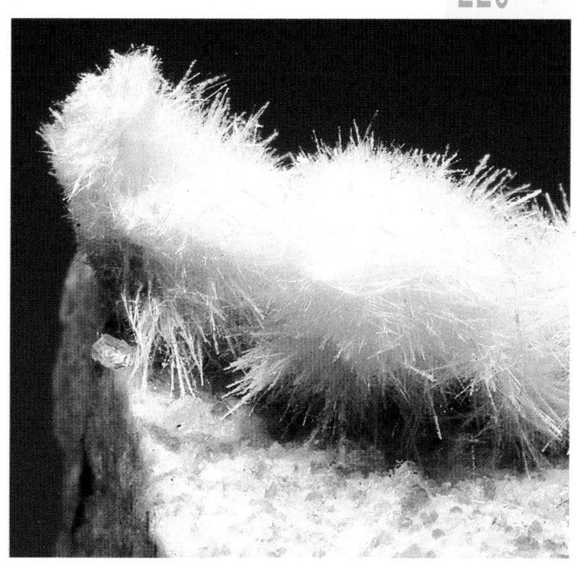

Formula $Ca_4Be_2Al_2Si_9O_{16}(OH)_2$
System Orthorhombic
Habit Forms fan-shaped globular aggregates, and composed of fibrous-radiating acicular crystals; white or colorless, with a vitreous luster; also present in distinct crystals, thin and tabular.
Environment Forms in the cavities of granitic pegmatites and in alpine-type hydrothermal veins.
Name and notes Named after Baveno, Italy, where the mineral was discovered.
Crystals from Beura, Piedmont, Italy.

Bustamite

230

Formula $(Ca,Mn^{2+})_3Si_3O_9$
System Triclinic
Habit Forms prismatic, tabular and acicular crystals; pink or brown-red, with a vitreous luster; also present in fibrous and granular aggregates.
Environment Forms in metamorphic skarn rich in minerals of manganese.
Name and notes Named after the Mexican general Anatasio Bustamenente (1780–1835), who discovered this mineral.

Group of crystals with a fibrous habit from Sweden.

Epididymite

231

5.5 2.55

Formula $NaBeSi_3O_7(OH)$
System Orthorhombic
Habit Forms pseudohexagonal tabular crystals, often star-shaped twins; white or colorless, with a vitreous to pearly luster.
Environment A rare mineral found in alkaline granitic and syenitic pegmatites.
Name and notes From the Greek *epi*, "similar," and *didymos*, "twinned," because its crystals resemble eudidymite.

Colorless crystals of epididymite associated with aegirine from Malawi.

Eudidymite

232

6 | 2.55

Formula $NaBeSi_3O_7(OH)$
System Monoclinic
Habit Forms tabular or lamellar crystals; white or colorless, with a vitreous to pearly luster.
Environment Even rarer than epididymite, forms in alkaline granitic pegmatites.
Name and notes From the Greek *eu*, "well," and *didymos*, "twinned," alluding to its typical twinned crystals.

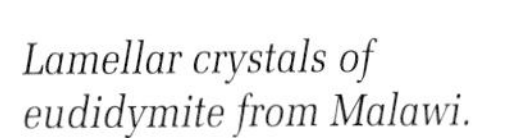

Lamellar crystals of eudidymite from Malawi.

Pectolite

233

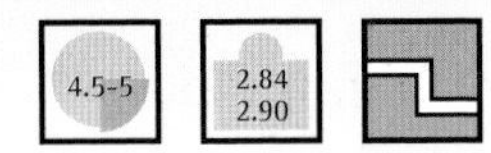

4.5–5 | 2.84 2.90

Formula $NaCa_2Si_3O_8(OH)$
System Triclinic
Habit Forms tabular to prismatic crystals and fibrous-radiating aggregates; white, whitish gray or pale yellow, with a silky luster.
Environment Present in intrusive igneous rocks such as nepheline syenites, in the cavities of effusive rocks such as basalts and andesites, and in diabasic veins; also present in metamorphic rocks such as serpentinites and skarn.
Name and notes From the Greek *pektos*, meaning "compacted," referring to its compact structure.

Group of fibrous-radiating crystals of pectolite from a basalt cavity at Tierno, near Mori, in the Trentino-Alto Adige, Italy.

Rhodonite

234

5.5
6.5

3.57
3.76

Formula $CaMn_4Si_5O_{15}$
System Triclinic
Habit Forms tabular and laminar crystals, commonly with rounded edges; pink, red, brownish red, yellow or black (through oxidation), with a vitreous luster; also common in massive or granular forms.
Environment Occurs in metamorphic skarn rich in minerals of manganese.
Name and notes From the Greek *rhodon,* "rose," in allusion to its characteristic color.

Right: Specimen of isolated crystals of rhodonite from Sweden.

Below: Specimen of rhodonite with galena from Broken Hill, Australia.

Serandite

235

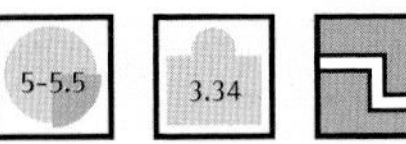

Formula $NaMn_2Si_3O_8(OH)$
System Triclinic
Habit Forms elongate prismatic and acicular crystals, and aggregates of radiating crystals, tabular or flat; red-pink, orange, salmon-pink, pale pink or reddish brown, with a vitreous luster.
Environment Occurs in alkaline igneous rocks, in particular sodalite and nepheline syenites, pegmatites and phonolites.
Name and notes Named after West African mineral collector Jean Sérand. First described at Rouma Island, Guinea.

Specimen of serandite with analcime and natrolite from Mont Saint-Hilaire, Quebec, Canada.

Wollastonite

236

4.5-5
2.86
3.09

Formula $CaSiO_3$
System Monoclinic and triclinic
Habit Forms prismatic and tabular crystals, commonly in fibrous aggregates; white, brown, yellow, pale green or colorless, with a vitreous to pearly luster.
Environment Common in metamorphic skarn rich in calcium; also present in alkaline igneous rocks and carbonatites.
Name and notes Named after British chemist and mineralogist William Wollaston (1766–1828).

Prismatic crystals of wollastonite from volcanic ejecta of Mount Somma-Vesuvius, Italy.

CHLORITE GROUP

This group includes nearly a dozen silicates that share very similar chemical, crystallographic and morphological characteristics. The various minerals in this group are difficult to identify or tell apart without detailed mineralogical analysis; some of them are quite widespread as components of metamorphic and igneous rocks.

Crystals of "kammerite," a chromiferous variety of clinochlore, from Turkey.

Clinochlore

2-2.5

2.60
3.02

Formula
$(Mg,Al)_6(Si,Al)_4O_{10}(OH)_8$
System Monoclinic
Habit Forms thin lamellar crystals with pseudohexagonal contours, generally joined in bundles; grass-green or dark green, with a pearly luster; also common in foliate, fibrous or granular aggregates.
Environment Common product of hydrothermal alteration of amphiboles, pyroxenes and biotites; component of metamorphic rocks and accessory of ultramafic igneous rocks.
Name and notes From the Greek words *klinein,* "to slope," and *chloros,* "green," alluding to its optical properties and characteristic color.

Pseudohexagonal crystal of clinochlore from the Susa Valley, Piedmont, Italy.

Cookeite

Formula $LiAl_4(Si_3Al)O_{10}(OH)_8$
System Monoclinic
Habit Forms flattened laminar crystals with pseudohexagonal contours; white, yellowish green, pink or brown, with a pearly luster; also common in spheroid and fibrous-radiating aggregates.
Environment Formed through hydrothermal alteration, characteristic of granitic pegmatites rich in lithium.
Name and notes Named after Josiah Cooke (1794–1863), chemist and mineralogist at Harvard University, Massachusetts. First described at Hebron, Maine.

Top: Specimen of cookeite from Elba, Italy.

Above: Cookeite associated with prismatic crystals of elbaite, also from Elba.

KAOLINITE-SERPENTINE GROUP

This group comprises two series of minerals that have many crystallochemical analogies, but which also have very different morphological characteristics. The kaolinite group includes minerals with a claylike and microcrystalline appearance, while the serpentines include minerals that form fibrous and flexible laminar crystals, some of which belong to the family that includes asbestos.

Specimen of antigorite with fibrous crystals.

Antigorite

239

Formula $Mg_3Si_2O_5(OH)_4$
System Monoclinic
Habit Forms generally minute crystals, typically flat, elongate and flexible, joined in laminar aggregates; greenish white, green or bluish green, with a resinous or silky luster.
Environment Forms in veins or fibrous masses associated with metamorphic and ultramafic igneous rocks.
Name and notes Named for Antigorio, Piedmont, Italy, where the mineral was first found.

Thin vein of laminar antigorite included in serpentinite rock from the Czech Republic.

Kaolinite

240

Formula $Al_2Si_2O_5(OH)_4$
System Monoclinic
Habit Forms massive aggregates, clayey, composed of microscopic crystals with pseudohexagonal contours; white or pale brown, with a pearly to earthy luster.
Environment Forms following the transformation of numerous silicates containing aluminum in hydrothermal environments or through surficial alteration.
Name and notes Named after Kao-ling, a hill in the People's Republic of China that was an early source of the mineral.

Specimen of microcrystalline kaolinite from the Czech Republic.

Clinochrysotile

241

2.5 | 2.53

Formula $Mg_3Si_2O_5(OH)_4$
System Monoclinic
Habit Forms flexible, parallel, cylindrical fibers similar to asbestos; white, pale green or dark green, with a silky luster.
Environment Forms fibers and fibrous masses associated with metamorphic rocks and ultramafic igneous rocks.
Name and notes From the Greek *chrysos,* "gold," and *tilos,* "fiber," referring to its color and characteristic habit.

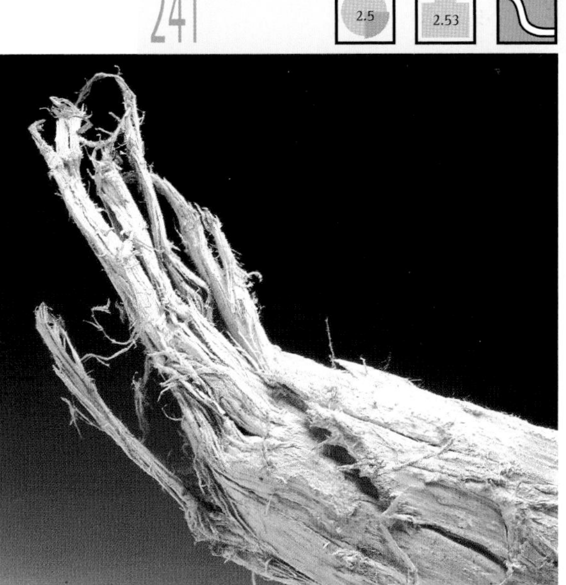

Specimen from the Malenco Valley, Lombardy, Italy.

Lizardite

2.5

2.55

Formula $Mg_3Si_2O_5(OH)_4$
System Hexagonal or trigonal
Habit Crystals are rare, with pseudohexagonal contours, joined in bunches; green, yellow or white, with a silky luster; also common in finely granulated and massive aggregates.
Environment Found in metamorphic rocks and ultramafic igneous rocks.
Name and notes Named after Lizard, in Cornwall, England, where the mineral was discovered. It is a trimorph with antigorite and clinochrysotile, with which it shares the same chemical composition but has a different crystal structure.

Above and below: Specimens of lizardite from the Apennines, near Piacenza, Italy.

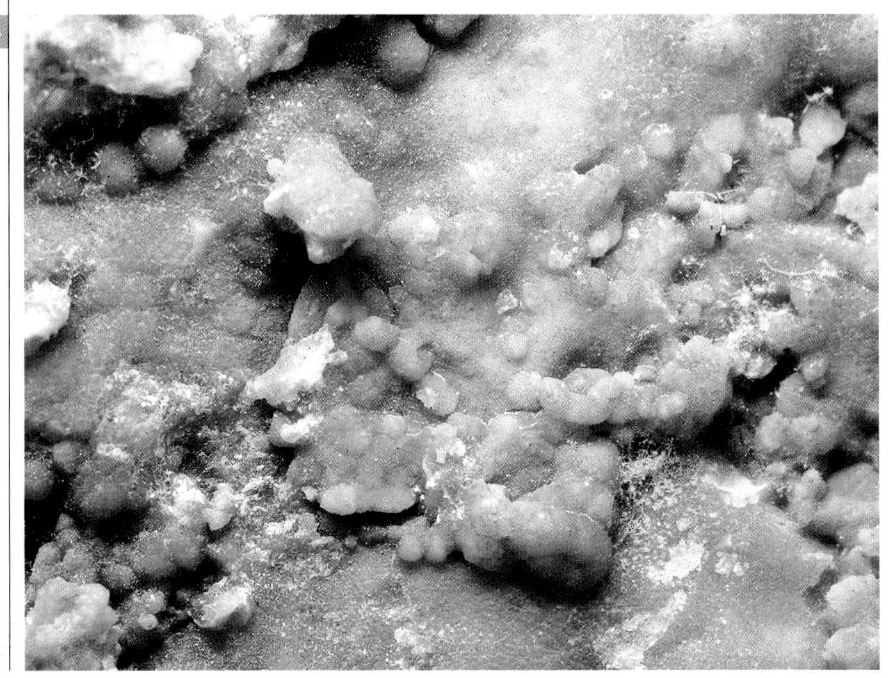

MICA GROUP

The minerals in this group (more than 40 are known) belong to the monoclinic crystal system, have highly complex formulas and are characterized by a perfect cleavage, high elasticity and flexibility. They are common in nature and are studied in petrology to understand the mechanism behind the formation of mica-bearing metamorphic and igneous rocks.

Phlogopite from Burma.

Biotite (series)

243

2.5-3 | 2.70 3.30

Crystal of biotite from Sri Lanka.

Formula $KMg_3AlSi_3O_{10}(OH,F)_2$ – $KFe_3AlSi_3O_{10}(OH,F)_2$
System Monoclinic

Habit Laminar crystals joined in bunches, prismatic with pseudohexagonal contours; dark green, black, reddish brown, yellow, brown-green or brown, with a vitreous to pearly luster.

Environment Component of metamorphic rocks (schist, gneiss, skarn) and intrusive igneous rocks, such as granites and syenites; less frequent in effusive magmatic rocks such as rhyolites and basalts.

Name and notes Named after French physicist and astronomer Jean Biot (1774–1862), who first studied its optical properties.

Phlogopite

244

2-3 | 2.78 2.85

Formula $KMg_3AlSi_3O_{10}(OH,F)$
System Monoclinic
Habit Forms lamellar to prismatic crystals joined in packets, tabular with pseudohexagonal outlines; reddish brown, brown-red, brown-yellow, brown, green or white, with a vitreous to pearly luster.
Environment Typical of metamorphic rocks rich in magnesium, such as marble and skarn; also present in ultramafic igneous rocks, in particular peridotites and kimberlites.
Name and notes From the Greek *phlogopos*, "fiery looking," alluding to its reddish coloration.

Crystal of phlogopite from the Kola Peninsula, Russia.

Lepidolite (series)

245

2.5-4 | 2.80 2.90

Formula $KLi_{1.5}Al_{1.5}AlSi_3O_{10}F_2 - KLi_2AlSi_4O_{10}F_2$
System Monoclinic
Habit Forms lamellar, tabular and prismatic crystals with pseudohexagonal outlines; pink, reddish pink or violet, with a vitreous to pearly luster.
Environment Occurs in granitic pegmatites rich in lithium.
Name and notes From the Greek *lepidos,* "scale," in reference to the foliate appearance characteristic of minerals in this series. Like biotite, new international mineralogical nomenclature identifies lepidolite as a term covering several minerals in the mica group.

Pink lamellar crystals of lepidolite from Pakistan.

Muscovite

246

2.5 | 2.77 2.88

Formula $KAl_2 AlSi_3O_{10}(OH)_2$
System Monoclinic
Habit Forms lamellar, tabular and prismatic crystals with pseudohexagonal outlines; gray, green-brown, yellow or red, with a vitreous to pearly luster.
Environment Component of metamorphic rocks (schist and gneiss) and of intrusive igneous rocks such as granites and pegmatites; also present with quartz veins in hydrothermal veins.
Name and notes Named for its use as a form of window glass (Muscovy glass) in Muscovy, an ancient province of Russia.

Group of lamellar crystals with pseudohexagonal outlines from Minas Gerais, Brazil.

Zinnwaldite (series)

247

2.5–4 | 2.90 3.02

Formula $KFe^{2+}{}_2Al_2Si_2O_{10}(OH)_2$ – $KLi_2AlSi_4O_{10}F_2$
System Monoclinic
Habit Forms lamellar, tabular and prismatic crystals with pseudohexagonal outlines, in some cases joined in rosettes; gray-brown, brown-yellow, green or pale violet, with a vitreous to pearly luster.
Environment Characteristic of high-temperature hydrothermal veins (greisen) found in granites and granitic pegmatites.
Name and notes Named after Zinnwald, Bohemia, Czech Republic, where the mineral was first described. New international mineralogical nomenclature identifies zinnwaldite as a term covering several minerals in the mica group.

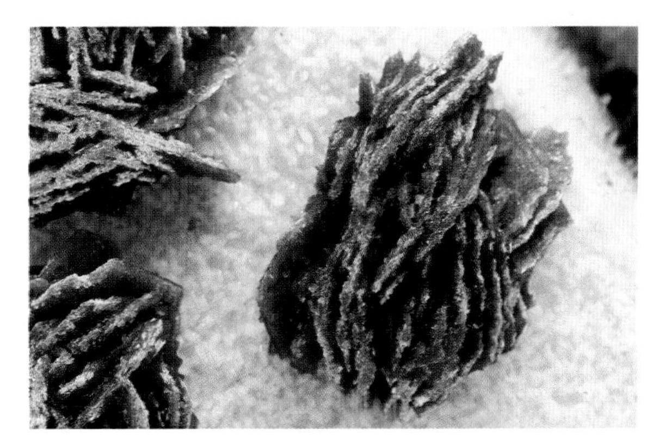

Rosette of crystals from Baveno, Piedmont, Italy.

Cavansite

248

3-4

2.21
2.31

Formula
$Ca(V^{4+}O)(Si_4O_{10})\cdot 4H_2O$
System Orthorhombic
Habit Forms characteristic spheroidal aggregates composed of elongate to prismatic crystals; bright sky-blue or greenish blue, with a vitreous luster.
Environment Found in cavities of basalts, basaltic breccia and andesitic tuffs.
Name and notes Named for its composition: calcium, vanadium and silicon. First identified near Chapman, Oregon.

Above and right: Aggregates of crystals of cavansite from the basalt quarry of Wagoli, near Poona, India.

Fluorapophyllite

249

4.5–5

2.33
2.37

Formula
$KCa_4Si_8O_{20}(F,OH) \cdot 8H_2O$
System Orthorhombic or tetragonal
Habit Forms beautiful tabular, prismatic or pseudocubic crystals; colorless, white, pink, yellow, greenish yellow or emerald-green, with a vitreous luster.
Environment This mineral of hydrothermal environments occurs in cavities in basalts and andesites, associated primarily with zeolites; also present in some metalliferous deposits.
Name and notes From *fluoro*, indicating the presence of that element, and the Greek *apo*, "distant," and *phyllazein*, "to produce leaves," as it cracks and exfoliates when heated.

Specimen from Rahuri, Maharashtra, India.

Hydroxyapophyllite

250

4.5–5

2.37

Formula
$KCa_4Si_8O_{20}(OH,F) \cdot 8H_2O$
System Orthorhombic or tetragonal
Habit Forms tabular to prismatic crystals; colorless, white, pink or greenish yellow, with a vitreous luster.
Environment Occurs in hydrothermal cavities in basalts and andesites; associated with zeolites.
Name and notes From the Greek *hydor*, "water," *apo*, "distant," and *phyllazein*, "to produce leaves," as this mineral cracks and flakes when heated.
Crystal of hydroxyapophyllite on gypsum from Malad, India.

Neptunite

251

5-6

3.19
3.23

Formula
$KNa_2Li(Fe,Mn)_2Ti_2Si_8O_{24}$
System Monoclinic
Habit Forms prismatic crystals with sharp terminations and square cross-section; black or brownish black, with a vitreous luster.
Environment Occurs in natrolite veins that cross serpentinite and schist; also present in some alkaline and granitic pegmatites.
Name and notes From Neptune, the Roman god of the sea; neptunite is associated with aegirine, named for Aegir, the Teutonic god of the sea.

Right and below: Crystals of neptunite displaying their characteristic prismatic form from San Benito County, California.

Okenite

252

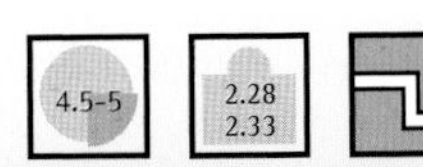

Formula $Ca_5Si_9O_{23} \cdot 9H_2O$
System Triclinic
Habit Not common in distinct laminar crystals, but rather in spheroidal aggregates composed of thin, fibrous and flexible crystals; white, with a silky luster.
Environment Occurs in the cavities of basalts and andesites; associated with zeolites.
Name and notes Named after German naturalist Lorenz Ocken (1779–1851). The name was originally written ockenite, but in 1830 it was changed to its current spelling by the mineralogist Hans von Kobell. First described on the island of Disko, Greenland.

Above and right: Spheroid aggregates of okenite from the Deccan Peninsula, India.

Petalite

253

6.5

2.41
2.42

Formula $LiAlSi_4O_{10}$
System Monoclinic
Habit Somewhat rare in tabular and prismatic crystals with striated faces, and more common in easily cleaved massive aggregates; colorless, white or gray, with a silky or vitreous luster.
Environment Characteristic of granitic pegmatites rich in lithium.
Name and notes From the Greek *petal,* "petal," in allusion to its characteristic way of cleaving. First found near Utö, Sweden.
Crystal of petalite from pegmatite veins, Elba, Italy.

Prehnite

6-6.5

2.80
2.95

Formula $Ca_2Al_2Si_3O_{20}(OH)_2$
System Orthorhombic
Habit Rare in distinct tabular crystals, and more common in globular aggregates formed of crystals with curving faces; green, white, yellow, gray or pink, with a vitreous luster.
Environment Characteristic of hydrothermal veins, associated with mafic igneous rocks and metamorphic rocks.
Name and notes Named for Colonel Hendrick von Prehn (1733–85), who discovered the mineral near the Cape of Good Hope, South Africa.

Aggregate of prehnite crystals with a globular habit from the basalt quarries of Poona, India.

Pyrophyllite

255

Formula $Al_2Si_4O_{10}(OH)_2$
System Monoclinic or triclinic
Habit Forms lamellar crystals united in radiating or spheroidal aggregates; white, yellow, apple-green, greenish gray or brownish green, with a pearly luster.
Environment Mineral of hydrothermal deposits, associated with schistose metamorphic rocks.
Name and notes From the Greek *pyr,* "fire," and *phyllon,* "leaf," since this mineral tends to exfoliate if heated enough.

Pyrophyllite from the Maira Valley, Piedmont, Italy.

Talc

256

1

2.58
2.83

Formula $Mg_3Si_4O_{10}(OH)_2$
System Monoclinic or triclinic
Habit Forms massive and globular aggregates and more rarely, flat crystals; pale green, white or brown, with a pearly luster.
Environment Component of metamorphic rocks, forms through hydrothermal alteration of igneous mafic rocks.
Name and notes Ancient term derived from the Arabic *talq,* meaning "pure," or "white," presumably alluding to the color of its powder.
Specimen of talc from the Trebbia Valley in the Apennines, Italy.

CANCRINITE GROUP

This group comprises about 15 silicates with hexagonal symmetry and complex formulas containing aluminum, calcium, potassium, sodium and anionic groups composed of carbon and sulfur. Cancrinite is the most common mineral in this group and is relatively widespread as the constituent of certain alkaline igneous rocks.

Specimen of afghanite from Sar-e-Sang, Badakshan, Afghanistan.

Afghanite

257

5.5–6 2.55

Formula $[(Na,K)_{22}Ca_{10}][Si_{24}Al_{24}O_{96}](SO_4)_6Cl_6$
System Trigonal
Habit Forms short prismatic and hexagonal crystals; blue or greenish blue, with a vitreous luster; also present in rounded granular aggregates.
Environment Present in metamorphic skarn rich in alkalis.
Name and notes Named after Afghanistan, where it was first found at Sar-e-Sang in Badakhshan.

Hexagonal crystal from Afghanistan.

Cancrinite

5-6

2.42
2.51

Formula $[(Ca,Na)_6(CO_3)_{1\text{-}1.7}][Na_2(H_2O)_2](Si_6Al_6O_{24})$
System Hexagonal
Habit Forms prismatic crystals terminating in hexagonal bipyramids; colorless, white, blue, grayish blue, honey-yellow or reddish orange, with a vitreous luster.
Environment Accessory of some alkaline igneous rocks; also present in alkaline metamorphic skarn.
Name and notes Named after Russian finance minister Count Georg Cancrin (Egor Kankrin) (1774–1845).

Right: Isolated hexagonal prismatic crystal with flat termination.
Below: Group of crystals, some of them twinned.

FELDSPAR GROUP

The silicates in this group (around 20 species have been recognized) constitute the most widespread family of minerals, found in a large number of rocks. They have triclinic or monoclinic symmetry and general formulas of the type XZ_4O_8, in which X can contain barium, calcium, potassium and sodium, while Z can contain silicon, aluminum and boron.

Specimen of the adularia variety of orthoclase from Pakistan.

Albite

259

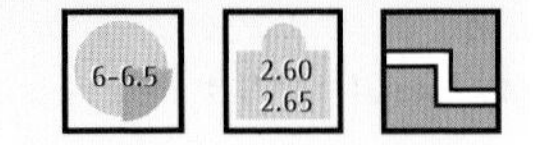

Tabular crystals of albite from Piedmont, Italy.

Formula $NaAlSi_3O_8$
System Triclinic
Habit Forms tabular crystals, possibly with curving faces, generally twinned, and joined in divergent aggregates; white, gray, pale green or colorless, with a vitreous luster; also common in granular or massive aggregates.
Environment Principal component of granites, pegmatites and other magmatic igneous rocks; also occurs in alpine-type hydrothermal veins and in gneissic metamorphic rocks.
Name and notes From the Latin *albus*, "white," in reference to its characteristic color; in the sodium-dominant end of the plagioclase series.

Anorthite

Formula $CaAl_2Si_2O_8$
System Triclinic
Habit Forms short pseudoprismatic crystals, frequently twinned; white, gray, reddish or colorless, with a vitreous luster; also common in granular and massive forms.
Environment Found in mafic igneous rocks; also present in metamorphic rocks, in particular granulites and skarn.
Name and notes From the Greek *an*, "not," and *orthos*, "right angle," since the crystals of this mineral have oblique faces. First described on Mount Somma-Vesuvius, near Naples, Italy. In the calcium-dominant end of the plagioclase series.

Crystal of anorthite with a pseudoprismatic habit from Mount Somma-Vesuvius, Italy.

Hyalophane

261

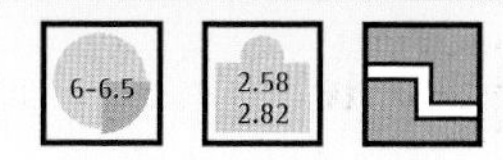

Group of clear prismatic crystals of hyalophane from the marble quarry of Legenbach in the Binn Valley, Switzerland.

Formula $(K,Ba)Al(Si_1Al)_3O_8$
System Monoclinic
Habit Forms prismatic crystals morphologically very similar to orthoclase; colorless or white, with a vitreous luster.
Environment Found in metamorphic deposits associated with minerals of barium and manganese.
Name and notes From the Greek *hyalos*, "glass," and *phainesthai*, "to seem," since the crystals of this mineral are always transparent. First described in the quarry of Legenbach in the Binn Valley, Switzerland.

6-6.5

2.54
2.57

Microcline

this mineral deviate only slightly from 90°.

Because microcline is indistinguishable from orthoclase, specific crystallochemical analyses are required to correctly identify the two minerals.

Bluish green pseudoprismatic crystals of microcline from granitic pegmatites at Pikes Peak, Colorado.

Microcline from granite pegmatites in Japan.

Formula $KAlSi_3O_8$
System Triclinic
Habit Forms pseudoprismatic crystals, commonly twinned, very similar to orthoclase; white, yellow, pale brown, reddish, green, blue or colorless, with a vitreous luster; also common in cleavable masses and in granular and massive aggregates.
Environment Component of many intrusive igneous rocks, in particular granites, granodiorites, pegmatites and aplites; also present in gneissic and amphibolic-rich metamorphic rocks; also found in alpine-type hydrothermal veins.
Name and notes From the Greek words *mikro*, "small," and *klinein*, "to incline," since the cleavage planes of

Orthoclase

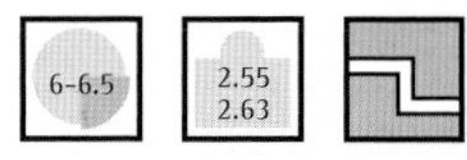

Formula $(K,Na)AlSi_3O_8$
System Monoclinic
Habit Elongate prismatic crystals, rough, tabular and commonly twinned; white, gray, pale yellow, red, green or colorless, with a vitreous luster; also common in cleavable masses and granular and massive aggregates.
Environment Component of many intrusive mafic rocks and of pegmatites; also present in gneissic and amphibolic-rich metamorphic rocks; also found in alpine-type hydrothermal veins.
Name and notes From the Greek *orthos*, "right angle," and *klas*, "to break," since its planes of cleavage intersect at nearly 90°.

Specimen of orthoclase from Elba, Italy.

Sanidine

264

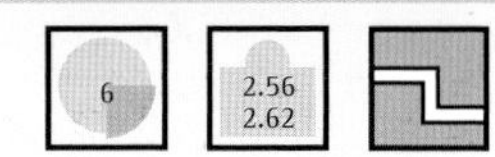

Formula $(K,Na)[(Si,Al)_4O_8]$
System Monoclinic
Habit Forms characteristic tabular crystals with square cross-section, commonly twinned; white, red or colorless, with a vitreous luster.
Environment Component of effusive igneous rocks; also present in metamorphic skarn rich in potassium.
Name and notes From the Greek *sanida*, "table," in reference to the tabular habit of its crystals.

Specimen of sanidine from the volcanic ejecta of Mount Somma-Vesuvius, Italy.

SCAPOLITE GROUP

This group consists of two silicates with tetragonal symmetry that form a completely miscible isomorphic series. The two end-members are marialite, which is sodium-dominant, and meionite, which is calcium-dominant. They are important phases because of their ability to store volatile constituents in the deep crust.

Crystal of marialite from Madagascar.

Marialite

265

Prismatic crystal from Finland.

Formula $Na_4Al_3Si_9O_{24}Cl$
System Tetragonal
Habit Forms tetragonal crystals terminating in poorly developed bipyramids; colorless, white, gray, pink, violet, blue, yellow, pale green, brown or orange-brown, with a vitreous or resinous luster.
Environment Characteristic of metamorphic rocks, in particular marble, skarn rich in sodium, granulites, gneiss and green schist; also present in mafic igneous rocks, pegmatites and alpine-type hydrothermal veins.
Name and notes Named after Maria Rosa, wife of the German mineralogist Gerhard von Rath (1830–88).

5-6

2.74
2.78

Meionite

Formula $Ca_4Al_6Si_6O_{24}CO_3$
System Tetragonal
Habit Forms tetragonal crystals terminating in poorly formed bipyramids; colorless, white, gray, pink, violet, blue, yellow, pale green, brown or orange-brown, with a vitreous or resinous luster.
Environment Characteristic of metamorphic rocks, in particular skarn rich in sodium, granulites, gneiss and green schist; also present in mafic igneous rocks.
Name and notes From the Greek *meion,* "less," because of the poorly formed terminations of its crystals.

Top: Crystal of meionite contained in volcanic ejecta from Mount Somma-Vesuvius, near Naples. Above: Detail showing the characteristic tetragonal prismatic habit of meionite crystals.

SODALITE GROUP

The four species of the sodalite group are silicates with cubic symmetry and highly complex formulas containing aluminum, calcium, sodium and anions such as chlorine and sulfur. It is interesting to note that the ornamental stone known as lapis lazuli is in fact composed of several members of this group.

Sodalite crystal from Mount Somma-Vesuvius, near Naples, Italy.

Haüyne

267

Formula $Na_6Ca_2Al_6Si_6O_{24}(SO_4)_2$
System Cubic

Habit Typical rhombododecahedral, cubic or octahedral crystals; bright blue, pale blue, greenish blue, white, gray, brown, green, yellow-green or colorless, with a vitreous luster; also present in granular and massive forms.
Environment Characteristic of igneous rocks and metamorphic skarn rich in alkalis.
Name and notes Named for the Abbé René-Just Haüy (1743–1822), founder of crystallography and Professor of Mineralogy at the Museum of Natural History in Paris.

Haüyne crystal with octahedral habit from the caldera of Sacrofano, Latium, Italy.

Lazurite

Formula $(Na,Ca)_8Si_6Al_6O_{24}[(SO_4),S,Cl,(OH)]_2$
System Cubic
Habit Rare rhombododecahedral crystals; dark blue, sky-blue or greenish blue, with a vitreous luster; common in granular and massive forms.
Environment Characteristic of metamorphic skarn with a carbonate-rich composition and marble rich in alkalis.
Name and notes From the Latin *lazhuward,* "blue," alluding to its typical color. First described in the region of Badakshan, Afghanistan.

Rare example of lazurite from volcanic ejecta of Mount Somma-Vesuvius, near Naples, Italy.

Sodalite

269

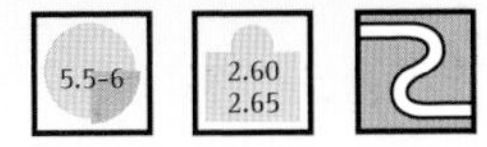

Formula $Na_8Al_6Si_6O_{24}Cl$
System Cubic
Habit Rare in distinct rhombododecahedral crystals; colorless, white, yellow, green, blue, violet or pink, with a vitreous luster.
Environment Found in igneous rocks rich in alkalis, in particular nepheline syenites and phonolites; also present in alkali-rich metamorphic skarn and marbles.
Name and notes Named after the mineral's sodium content.
Vitreous crystal of sodalite from volcanic ejecta of Mount Somma-Vesuvius, near Naples, Italy.

ZEOLITE GROUP

The nomenclature of this vast and important group of silicates was recently redefined. The members of this group have channels that contain molecules of water that can be expelled or replaced without changing the structure. For this reason, these minerals can behave as ion exchangers and be used, for example, as natural water filters in industrial or civic installations.

Specimen of natrolite from basalt cavities in India.

Analcime

270

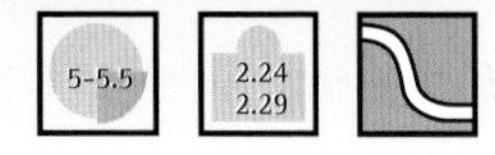

Perfect crystal of analcime from the Alps of Souse, Trentino-Alto Adige, Italy.

Formula $NaAlSi_2O_6 \cdot H_2O$
System Cubic
Habit Forms generally trapezohedral crystals; colorless, white, pink, gray, greenish or yellowish, with a vitreous luster; also common in massive and granular forms.
Environment Found in cavities of effusive igneous rocks, in particular basalts and phonolites; also present in veins of hydrothermal environments.
Name and notes From the Greek *annalist*, "weak," in allusion to its weak piezoelectricity when heated and rubbed. First described on Cyclops Island, Sicily.

Chabazite-Ca

271

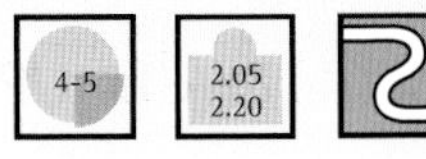

Formula
$Ca_2(Al_4Si_8O_{24})\cdot 13H_2O$
System Trigonal
Habit Present in a large variety of crystalline habits, including pseudocubic, rhombohedral and tabular, and pseudohexagonal through twinning; white, yellow, pink, red or colorless, with a vitreous luster.
Environment Forms in cavities of volcanic rocks such as basalts and andesites; also present in granites and granitic pegmatites, in veins of hydrothermal environments, and in skarn assemblages.
Name and notes From the Greek *chabazios,* "melody," one of the 20 stones named in the poem *Peri Lithos,* ascribed to the mythical Greek poet Orpheus. Chabazite-Na and chabazite-K were recently recognized as distinct species.
Rhombohedral crystal from India.

Gismondine

272

4.5 | 2.20 2.26

Formula
$Ca_4(Al_8Si_8O_{32})\cdot 16H_2O$
System Monoclinic
Habit Forms pseudobipyramidal crystals; white, grayish, bluish or colorless, with a vitreous luster.
Environment Characteristic of igneous rocks such as basalts and phonolites; also present in skarn assemblages.
Name and notes Named after Carlo Giuseppe Gismondi (1762–1824), who first recognized it as a new mineral species. First found at Capo di Bove, near Rome, Italy.
Specimen from the Zebrù Valley, Lombardy, Italy.

Gmelinite-Na

4.5

2.02
2.17

Formula
$Na_{7.5}(Al_{7.5}Si_{16.5}O_{48})\cdot21.5H_2O$
System Hexagonal
Habit Forms rhombohedral, pseudopyramidal and tabular crystals, with hexagonal contours, often twinned; white, reddish, salmon-pink, yellowish, greenish white or colorless, with a vitreous luster.
Environment Found in the cavities of volcanic rocks such as basalts and andesites.
Name and notes Named after the German chemist and mineralogist Christian Gmelin (1792–1860). First found at Montecchio Maggiore in the province of Vicenza, Italy. Gmelinite-CA and gmelinite-K have been recognized as distinct species.

Above and below: Specimens of gmelinite from Montecchio Maggiore, Italy.

Heulandite-Ca

3.5–4

2.10
2.20

Formula $(Ca_{0.5},Na,K)_9[(Al_9Si_{27}O_{72}]\cdot 24H_2O$
System Monoclinic
Habit Forms very characteristic elongate and tabular crystals; colorless, white, yellow, red, pink, orange and brown, with a vitreous or pearly luster.
Environment Forms in cavities of volcanic rocks, such as basalts and andesites; also present in some granite and granitic pegmatites.
Name and notes Named after British mineral collector John Heuland (1778–1856). Heulandite-Sr, huelandite-Na, and huelandite-K have been recognized as distinct species.

Group of crystals from Iceland.

Laumontite

275

3–4

2.23
2.41

Formula $Ca_4[Al_8Si_{16}O_{48}]\cdot 18H_2O$
System Monoclinic
Habit Forms prismatic crystals with square cross-section and sharp terminations; white, gray, pale yellow, pink or yellowish brown, with a vitreous luster; also common in fibrous-radiating aggregates.
Environment Mineral of hydrothermal environments, develops in cavities of igneous rocks and as a filling of fractures in metamorphic rocks.
Name and notes Named after French mineralogist François de Laumont (1747–1834), who discovered it at Huelgoet, Brittany, France. It loses water on exposure to air and readily converts to a whitish powder.

Specimen from the Val Formazza, Piedmont, Italy.

Leucite

276

Magnificent pseudo-isometric crystal from Ariccia, Latium, Italy.

Formula $KAlSi_2O_6$
System Tetragonal
Habit Forms pseudocubic and pseudo-isometric crystals with typical lengthwise striations on the faces as a manifestation of twinning; white, gray or colorless, with a vitreous luster; also common in granular and massive aggregates.
Environment Characteristic of effusive igneous rocks rich in potassium.
Name and notes From the Greek *leukos,* "white," in reference to its characteristic color. First described on Mount Somma-Vesuvius, near Naples, Italy.

Mesolite

277

Formula
$Na_2Ca_2(Al_6Si_9O_{30})\cdot 8H_2O$
System Orthorhombic
Habit Forms very narrow acicular to prismatic crystals, in some cases capillary, joined in radiating aggregates; colorless, white, gray or yellowish, with a vitreous luster, or silky luster if the mineral has a fibrous habit.
Environment Found in cavities of volcanic rocks, such as basalts and andesites, in hydrothermal veins, and in some skarn assemblages.
Name and notes From the Greek *mesos,* "half," and *lithos,* "stone," in reference to its intermediate characteristics between natrolite and scolecite. First described in Nova Scotia, Canada.

Mesolite found in cavities of exposed basalts near Montresta, Sardinia, Italy.

Natrolite

5-5.5

2.20
2.26

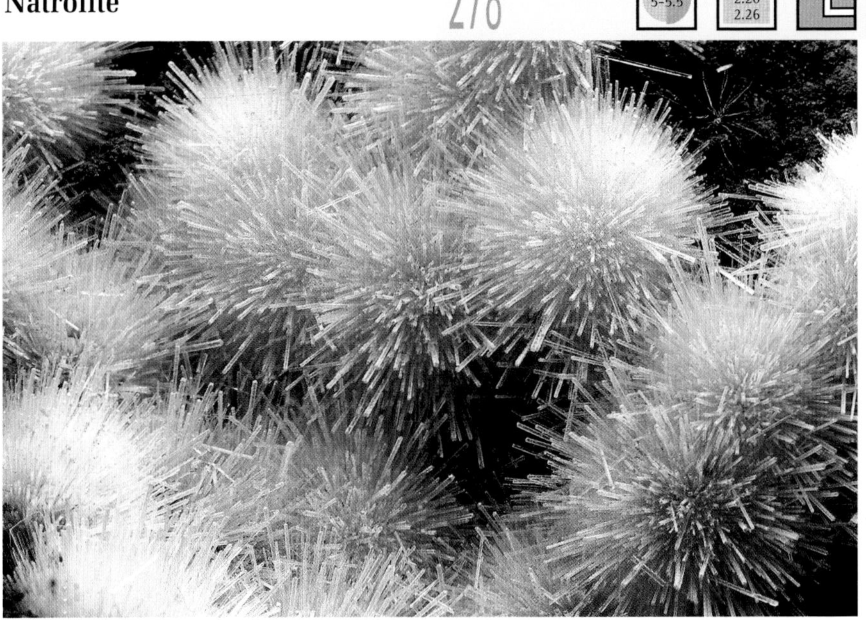

Formula $Na_2Al_2Si_3O_{10}·2H_2O$
System Orthorhombic
Habit Forms elongate or short prismatic crystals, striated, terminating in bipyramidal faces; colorless, white, gray, yellow or pink, with a vitreous luster, or a silky one if the mineral has a fibrous-radiating habit.
Environment Found in the cavities of volcanic rocks such as basalts and andesites; also found in some syenites and granitic alkaline pegmatites.
Name and notes From the Greek words *nitron,* "niter," and *lithos,* "stone," referring to its chemical composition. First described at Hegau, Germany.

Above: Specimen of natrolite from Altavilla, Italy.

Below: Specimen from Arizona.

Phillipsite-K

279

4-4.5 2.20

Formula $(K,Ca_{0.5},Na,Mg_{0.5},Sr_{0.5})[Al_9Si_{27}O_{72}]\cdot 24H_2O$
System Monoclinic
Habit Forms prismatic crystals, generally cross-shaped twins, joined in aggregates; colorless, white or yellowish, with a vitreous luster.
Environment Typical of basaltic cavities, also present in evaporite and carbonate-rich sedimentary rocks.
Name and notes Named after mineralogist William Phillips (1775–1828), founder of the Geological Society of London. First found near Aci Castello, Sicily in Italy. Phillipsite-Na and phillipsite-Ca have been recognized as distinct species.

Crystals from Mount Somma-Vesuvius, near Naples, Italy.

Pollucite

280

6.5-7 2.68 3.03

Formula $(Cs,Na)[AlSi_2O_6]\cdot \underline{n}H_2O$
System Cubic
Habit Forms somewhat rare cubic crystals, icositetrahedral and rhombododecahedral; colorless, white, gray or pink, with a vitreous luster; more common in granular and massive aggregates.
Environment Characteristic of granitic pegmatites rich in lithium.
Name and notes From the Latin *Pollux,* twin brother of Castor, sons of Zeus and heroes in Greek mythology, alluding to its intimate relationship to the mineral "castorite" (later renamed petalite).

Rare and perfect crystal of pollucite from Pakistan.

Scolecite

Formula $CaAl_2Si_3O_{10}·3H_2O$
System Monoclinic
Habit Forms elongate, acicular and prismatic crystals, with square cross-section; colorless or white, with a vitreous luster, or a silky luster if the mineral is fibrous.
Environment Forms in cavities of basalts, in syenitic veins and gabbros; also present in some metamorphic rocks, in particular gneiss and amphibolites.
Name and notes From the Greek *skolex*, "worm," because if put to the flame the borax that covers the mineral melts and tends to form vermiform curls.

Group of elongate and prismatic colorless crystals of scolecite associated with white crystals of hydroxy-apophyllite from basaltic cavities near Malad, India.

Stilbite-Ca

282

3.5–4 | 2.18 2.20

Specimen from India.

Formula $(Ca_{0.5},K,Na)_9[Al_9Si_{27}O_{72}]·28H_2O$
System Monoclinic
Habit Forms thin tabular crystals, joined in divergent aggregates, globular; white, yellowish, gray, pink, reddish, orange, brown or colorless, with a vitreous luster.
Environment Forms in cavities of volcanic rocks such as basalts and andesites, and in some granites and granitic pegmatites; also present in metamorphic rocks and in alpine-type hydrothermal veins.
Name and notes From the Greek *stilbein*, "to glitter," or from *stilbe*, "mirror," in reference to this mineral's glassy sheen. Stilbite-Na has been recognized as a distinct species.

Thomsonite-Ca

283

5–5.5 | 2.23 2.39

Formula $NaCaAl_5Si_5O_2 \cdot 6H_2O$
System Orthorhombic
Habit Forms thin tabular and prismatic crystals and spherical aggregates, globular; white, pale yellow, brown, green or colorless, with a vitreous luster.
Environment Forms in cavities of volcanic rocks such as basalts and andesites; also present in some skarn assemblages.
Name and notes Named after Thomas Thomson (1773–1852), Professor of Chemistry at the University of Glasgow, Scotland, who was the first to study this mineral. First described at Dumbartonshire, Scotland.

Crystals from Mount Somma-Vesuvius, near Naples, Italy.

Yugawaralite

284

4.5–5 | 2.20 2.23

Formula $Ca_2[Al_4Si_{12}O_{32}] \cdot 8H_2O$
System Monoclinic
Habit Forms flat tabular crystals; colorless or white, with a vitreous luster.
Environment Forms in cavities of volcanic rocks, such as basalts and andesites, and in quartz veins of hydrothermal deposits.
Name and notes Named for Yugawara, in the Honshu region of Japan, where the mineral was discovered.

Group of flat tabular crystals from basalts in India.

Danburite

285

7-7.25 | 2.93 3.02

Formula $Ca[B_2Si_2O_8]$
System Orthorhombic
Habit Forms prismatic or rhombohedral crystals with square cross-section; white, yellow, yellowish brown, greenish or colorless, with a vitreous to resinous luster; also common in massive aggregates.
Environment Forms in granites, granitic pegmatites and skarn assemblages rich in calcium and boron.
Name and notes Named after Danbury, Connecticut.

Danburite specimen from Mexico, where it was first found, in 1839.

Nepheline

286

5.5-6 | 2.55 2.66

Typical crystal of nepheline with a hexagonal habit from Mount Somma-Vesuvius, Italy.

Formula $(Na,K)AlSiO_4$
System Hexagonal
Habit Forms short hexagonal prismatic crystals with flat terminations; colorless, white, gray or yellowish, with a vitreous luster; also frequent in granular and massive aggregates.
Environment Characteristic mineral of alkaline rocks, such as nepheline syenites, phonolites, gabbros and skarn assemblages.
Name and notes From the Greek *nephele,* "cloud," alluding to its cloudy color when immersed in strong acid. First described on Mount Somma-Vesuvius, Italy.

ORGANIC MINERALS

Class 10

Mellite

287

2-2.5

1.55
1.65

Formula $Al_2[C_6(COO)_6]\cdot16H_2O$
System Tetragonal
Habit Somewhat rarely forms crystals, prismatic or pyramidal; honey-yellow, reddish brown or white, with a resinous luster; common in massive aggregates.
Environment Secondary mineral associated with deposits of carbon and lignite (coal).
Name and notes From the Greek *meli,* "honey," alluding to its characteristic color. First found in the locality of Arten in Thuringia, Germany.

Right and below: Specimens of typical bipyramidal crystals of mellite from Hungary.

Whewellite

2.5–3

2.22

Formula $CaC_2O_4 \cdot H_2O$
System Monoclinic
Habit Forms beautiful short prismatic crystals, commonly heart-shaped twins; colorless, yellow or brown, with a vitreous luster.
Environment Secondary mineral associated with deposits of carbon and lignite; also forms in hydrothermal environments and in metalliferous deposits.
Name and notes Named after British naturalist and scientist William Whewell (1794–1866), inventor of the popular crystallographic system that was later developed by his student William Miller. The mineral was first studied at Havre, Montana.

Top: Crystals of whewellite from Kladno, Czech Republic.

Above: Crystals from Bilina, Czech Republic.

REFERENCES

ESSENTIAL GLOSSARY

alkaline rocks or minerals rich in alkali metals, such as potassium and sodium.
amphibolite metamorphic rock containing amphiboles and Na-Ca feldspar (plagioclase).
aplite intrusive igneous vein rock with a fine grain-size and granitic composition.
basalt effusive igneous rock containing pyroxenes, plagioclase and olivine.
carbonatite effusive igneous rock containing more than 50 percent carbonates.
dacite effusive igneous rock containing potassium-rich feldspar, plagioclase, quartz, pyroxenes and biotite.
diorite intrusive igneous rock containing plagioclase, quartz, amphiboles and biotite.
eclogite metamorphic rock containing sodic pyroxene (omphacite) and garnet.
gabbro intrusive igneous rock containing feldspars, amphiboles and pyroxenes.
gneiss metamorphic rock composed of segregations of feldspars and quartz, and of biotite and other minerals.
granulite metamorphic rock containing feldspars and pyroxenes.
hornfels fine-grained metamorphic rock composed of high-temperature silicates.
hydrothermal pertaining to the circulation of hot aqueous solutions or gases within the earth's crust.
mafic, ultramafic synonyms of basic and ultrabasic, terms that indicate rocks poor in silicon oxide, with little or no quartz and rich in amphiboles, pyroxenes, mica, olivine and other silicates.
marble metamorphic rock predominantly composed of carbonates (calcite, dolomite).
monzonite intrusive igneous rock containing potassium-rich feldspar, plagioclase and amphiboles.
pegmatite intrusive igneous vine rock, very coarse-grained, usually with granitic composition.
peridotite ultramafic rock containing pyroxenes and olivine.
phonolite effusive alkaline igneous rock containing potassium-rich feldspar, pyroxenes and silicates poor in silica.
plagioclase very common feldspar series where composition can vary from albite to anorthite.
rhyolite effusive igneous rock containing quartz, potassium-rich feldspar, calcium-rich plagioclase, amphiboles and micas.
rodingite metamorphic vein rock containing calcium silicates.
schist metamorphic rock characterized by planar surfaces (foliations); named according to the prevailing mineral, as in chlorite schist, mica schist, talc schist.
serpentinite ultramafic metamorphic rock primarily composed of minerals of the serpentine group.
skarn metamorphic-metasomatic rocks composed of calc-silicates, iron and magnesium. It forms on contact with plutons and involves the transformation of carbonate-rich rocks into which significant quantities of silica, aluminum, iron and magnesium are introduced.
syenite intrusive igneous rock containing potassium-rich feldspar, sodium plagioclase, little or no free quartz and amphiboles.

CLASSIFICATION OF MINERALS

NATIVE ELEMENTS
Antimony
Arsenic
Bismuth
Copper
Diamond
Gold
Graphite
Mercury
Platinum
Silver
Sulfur

SULFIDES

Pyrite group
Hauerite
Pyrite
Sperrylite

Acanthite
Arsenopyrite
Bismuthinite
Bornite
Calaverite
Carrollite
Chalcocite
Chalcopyrite
Cinnabar
Covellite
Enargite
Galena
Greenockite
Kermesite
Marcasite
Millerite
Molybdenite
Orpiment
Pyrrhotite
Realgar
Skutterudite
Sphalerite
Stibnite
Wurtzite

SULFOSALTS

Tetrahedrite group
Tennantite
Tetrahedrite

Boulangerite
Bournonite
Proustite
Pyrargyrite

HALIDES
Atacamite
Boleite
Chlorargyrite
Cryolite
Fluorite
Halite
Sylvite
Villiaumite

OXIDES

Hematite group
Corundum
Hematite

Perovskite group
Loparite-(Ce)
Perovskite

Pyrochlore group
Betafite
Microlite
Pyrochlore

Rutile group
Cassiterite
Plattnerite
Pyrolusite
Rutile

Spinel group
Chromite
Franklinite
Gahnite
Magnetite
Spinel

Aeschynite-(Y)
Anatase
Bixbyite
Brookite
Chrysoberyl
Cuprite
Euxenite-(Y)
Ferberite
Ferrocolumbite – Manganocolumbite
Hübnerite
Ilmenite
Manganotantalite
Opal
Quartz

Thorianite
Uraninite

HYDROXIDES
Brucite
Becquerelite
Diaspore
Goethite
Manganite

CARBONATES

Aragonite group
Aragonite
Cerussite
Strontianite
Witherite

Calcite group
Calcite
Magnesite
Rhodochrosite
Siderite
Smithsonite

Dolomite group
Ankerite
Dolomite
Kutnohorite

Artinite
Aurichalcite
Azurite
Hydrozincite
Malachite
Parisite-(Ce)
Phosgenite

BORATES
Behierite
Boracite
Canavesite
Colemanite
Hambergite
Jeremejevite
Rhodizite

SULFATES

Barite group
Anglesite
Barite
Celestine

Ettringite group
Ettringite
Sturmanite
Thaumasite

Alunite
Anhydrite
Brochantite
Chalcanthite
Cyanotrichite
Gypsum
Linarite
Voltaite

CHROMATES, MOLYBDATES, TUNGSTATES
Crocoite
Powellite
Scheelite
Stolzite
Wulfenite

PHOSPHATES, ARSENATES, VANADATES

Apatite group
Fluorapatite
Hydroxylapatite
Mimetite
Pyromorphite
Vanadinite

Autunite group
Autunite
Torbernite

Vivianite group
Erythrite
Vivianite

Adamite
Amblygonite – Montebrasite
Beryllonite
Brazilianite
Cafarsite
Descloizite
Herderite – Hydroxyl-herderite
Lazulite – Scorzalite
Legrandite
Liroconite
Lithiophilite – Triphylite
Ludlamite
Monazite-(Ce)
Pharmacosiderite
Phosphophyllite
Wardite
Wavellite
Xenotime-(Y)

SILICATES

NESOSILICATES

Gadolinite group
Datolite
Gadolinite-(Y)
Hingganite-(Y)

Garnet group
Almandine
Andradite
Grossular
Pyrope
Spessartine
Uvarovite

Olivine group
Fayalite
Forsterite

Andalusite
Braunite
Euclase
Kyanite
Phenakite
Sapphirine
Sillimanite
Staurolite
Titanite
Topaz
Uranophane
Zircon

SOROSILICATES

Axinite group
Ferro-axinite
Manganaxinite
Tinzenite

Epidote group
Allanite-(Y)
Epidote
Piemontite
Zoisite

Barylite
Bertrandite
Chevkinite-(Ce)
Hemimorphite
Ilvaite
Thortveitite
Vesuvianite

CYCLOSILICATES

Osumilite group
Milarite
Osumilite

Tourmaline group
Dravite
Elbaite
Liddicoatite
Schorl
Uvite

Benitoite
Beryl
Cordierite
Dioptase
Eudialyte

INOSILICATES

Amphibole group
Actinolite
Anthophyllite
Arfvedsonite
Ferrohornblende
Glaucophane
Kaersutite
Pargasite
Riebeckite
Tremolite

Monoclinic pyroxene group
Aegirine
Augite
Diopside
Hedenbergite
Jadeite
Johannsenite
Omphacite
Spodumene

Babingtonite
Bavenite
Bustamite
Epididymite
Eudidymite
Pectolite
Rhodonite
Serandite
Wollastonite

PHYLLOSILICATES

Chlorite group
Clinochlore
Cookeite

Kaolinite-serpentine group
Antigorite
Clinochrysotile
Kaolinite
Lizardite

Mica group
Biotite (series)
Lepidolite (series)
Muscovite
Phlogopite
Zinnwaldite (series)

Cavansite
Fluorapophyllite
Hydroxyapophyllite
Neptunite
Okenite
Petalite
Prehnite
Pyrophyllite
Talc

TECTOSILICATES

Cancrinite group
Afghanite
Cancrinite

Feldspar group
Albite
Anorthite
Hyalophane
Microcline
Orthoclase
Sanidine

Scapolite group
Marialite
Meionite

Sodalite group
Haüyne
Lazurite
Sodalite

Zeolite group
Analcime
Chabazite-Ca
Gismondine
Gmelinite-Na
Heulandite-Ca
Laumontite
Leucite
Mesolite
Natrolite
Phillipsite-K
Pollucite
Scolecite
Stilbite-Ca
Thomsonite-Ca
Yugawaralite

Danburite
Nepheline

ORGANIC MINERALS
Mellite
Whewellite

INDEX OF THE MINERALS

BIBLIOGRAPHY AND WEBSITES

Anthony, John W., et al.
Handbook of Mineralogy I, II, III, IV
Tucson: Mineral Data Publishing, 1997–2003

Bayliss, Peter
Glossary of Obsolete Mineral Names
Tucson: The Mineralogical Record, 2000

Bishop, A.C., et al.
Guide to Minerals, Rocks and Fossils
Richmond Hill, Ontario: Firefly Books, 2005

Blackburn, William H., and William H. Dennen
Encyclopedia of Mineral Names
Ottawa: The Canadian Mineralogist, 1997

Bloss, F. Donald
Crystallography and Crystal Chemistry
Mineralogical Society of America, 1994

Chesterman, Charles W.
The Audubon Society Field Guide to North American Rocks and Minerals
New York: Alfred A. Knopf, 1978

Gaines, Richard V., et al.
Dana's New Mineralogy, 8th Edition
New York: John Wiley & Sons, 1997

De Fourestier, Jeffrey
Glossary of Mineral Synonyms
Ottawa: The Canadian Mineralogist, 1998

Derr, W.A., R.A. Howie and J.Z. Zussman
The Rock-Forming Minerals
Longman Scientific and Technical, 1998

Klein, Cornelius, and Cornelius S. Hulburt, Jr.
Manual of Mineralogy, 21st edition
New York: John Wiley & Sons, 1993

Mandarino, Joseph A. and M.E. Back
Fleischer's Glossary of Mineral Species
Tucson: The Mineralogical Record, 2004

The Nomenclature of Minerals: A Compilation of IMA Reports
Ottawa: The Canadian Mineralogist, 1998

Oldershaw, Cally
Firefly Guide to Gems
Richmond Hill, Ontario: Firefly Books, 2004

Pezzotta, Federico
Madagascar, a Mineral and Gemstone Paradise
Lapis International, 1999

Putnis, Andrew
An Introduction to Mineral Sciences
New York: Cambridge University Press, 1992

Strunz, Hugo, and Ernest H. Nickel
Strunz Mineralogical Tables
Stuttgart, 2001

Teaching Mineralogy
Mineralogical Society of America, 1996

Internet websites

There are many journals devoted to mineralogy, most of them concerned with mineral collecting. Perhaps the best known in the field is *The Mineralogical Record*, which can be found on the web at www.minrec.org. The Canadian Mineralogical Association publishes the excellent technical journal *The Canadian Mineralogist*, with articles available for purchase and download online at pubs.nrc-cnrc.gc.ca/mineral/. The association's website is at www.mineralogicalassociation.ca. Lastly, *The American Mineralogist* has selected classic articles archived online at www.minsocam.org/MSA/AmMin/AmMineral.html.